International Perspectives on Hazardous Waste Management

International Perspectives on Hazardous Waste Management

A Report from the International Solid Wastes and Public Cleansing Association (ISWA) Working Group on Hazardous Wastes

edited by

WILLIAM S. FORESTER

*ISW/ISWA Secretariat,
American Public Works Association,
Chicago, USA*

and

JOHN H. SKINNER

*Chairman, ISWA Working Group on Hazardous Wastes
US Environmental Protection Agency,
Washington, USA*

1987

ACADEMIC PRESS

Harcourt Brace Jovanovich, Publishers

London Orlando San Diego New York
Austin Boston Sydney Tokyo Toronto

ACADEMIC PRESS INC. (LONDON) LTD.
24/28 Oval Road,
London NW1 7DX

United States Edition published by
ACADEMIC PRESS INC.
Orlando, Florida 32887

Copyright © 1987 by
ACADEMIC PRESS INC. (LONDON) LTD.

All rights reserved
No part of this publication may be reproduced or transmitted in any form
or by any means, electronic or mechanical, including photocopy,
recording, or any information storage and retrieval system, without
permission in writing from the publisher.

ISBN: 0.12.262165.4

Phototypeset by Burns & Smith, Derby.

Printed by St Edmundsbury Press,
Bury St Edmunds, England.

ISWA Working Group on Hazardous Waste

CHAIRMAN

John H. Skinner,
US Environmental Protection Agency,
Washington, DC 20460, USA

AUSTRIA

Dr Gerhard Vogel and Christoph Scharff,
Institut für Technologie und Warenwirtschafslehre der Wirtschaftsuniversitat,
Augasse 2-6,
A-1090 Vienna

DENMARK

Dr Ing. Klaus Muller,
National Agency of Environmental Protection,
Ministry of the Environment,
29, Strandgade,
DK-1401 Copenhagen K

FEDERAL REPUBLIC OF GERMANY

Dr Ing Gerhard Sierig,
Berliner Stadtreinigungsbetriebe
Ringbahnstrasse 96,
D 1000 Berlin 42

FRANCE

Jean-Bernard Leroy,
Ingenieur Civil de L-Aeronautique,
Compagnie Générale des Eaux,
52, rue D'Anjou,
75008 Paris

ITALY

Dr Alberto Piepoli,
SEVESO,
Regione Lombardia,
Via S. Carlo, n.4,
Milano

JAPAN

Dr Sachiho Naito,
Civil Engineering Department,
Kanto Gakuin University,
2-15-2, Oyamadai,
Setagaya-Ky, Tokyo 158

THE NETHERLANDS

Ir D. den Ouden,
AVR Chemie NV,
Postbus 1120,
3180 AC Rosenburg

SOUTHERN AFRICA

A. C. de Bruin and Stan Verrier,
Institute of Waste Management (Southern Africa),
P.O. Box 28,
Muizenberg 7945, South Africa

SPAIN AND PORTUGAL

Julian Uriarte Jaureguixar,
ATEGRUS, Secretario General,
Mugica y Butron, 10-2 -Dpto. 1,
48007 Bilbao

SWEDEN

Kenneth Andersson,
President,
Graab-Kemi,
Box 95,
401 21 Gothenburg 1

UNITED KINGDOM

Dr David C. Wilson,
Environmental Resources Ltd.,
106 Gloucester Place,
London W1H 3DB

UNITED STATES

William Forester (Secretary),
ISW/ISWA Secretariat,
American Public Works Association,
1313 East 60th Street,
Chicago, Illinois 60637

Preface

The Working Group on Hazardous Wastes was formed by action of the Governing Board of the International Solid Wastes and Public Cleansing Association (ISWA) at its annual business meeting, September 18, 1984, in Philadelphia, Pennsylvania, USA. Likewise approved was a three-year Work Plan that culminated this year with the publication of this report, and the ISWA-Japan Waste Management Association co-sponsored "1987 Hazardous Waste Symposium", held in Tokyo, June 22–24, 1987.

The working group held its first meeting in Milan, Italy, March 7, 1985, its membership comprised of 12 members and a secretary, who were selected by the 12 respective national member organizations of ISWA. An outline, from which the 12 country reports subsequently were produced, was approved. The working group met approximately twice each year to discuss progress and plan the next stage of report activity. At the request of the working group, the United Kingdom member, Dr David C. Wilson, Environmental Resources Limited, agreed to serve as consultant and to prepare the summary report that appears as Chapter 2 of this publication.

Appreciation must be extended first to the 12 members of the working group, to their agencies and firms, and to their national member organizations, for the work that produced the country reports, for moral support as well as assistance, and for travel funds that allowed our meetings to take place. Special appreciation is extended to Dr John H. Skinner, the US Environmental Protection Agency (EPA), for his leadership as chairman of the working group. Special thanks goes to the US EPA for grant monies that supported in part the preparation of the summary report.

Appreciation is also extended to three other international organizations working in the hazardous waste management field for review and comments that made this a better book than it otherwise would have been, although they in no way are responsible for its shortcomings. The working group met formally at least once during the past three years with officials of the Waste Management Policy Group of the Organization for Economic Cooperation and Development (OECD), the World Health Organization (WHO), and the United Nations Environmental Programme (UNEP).

Final thanks and appreciation must necessarily go to the officers and editors at Academic Press Inc. (London) Limited for their excellent assistance and consideration in the preparation of the manuscript of the book for publication within an abbreviated period of time. The working group would be remiss, indeed, if we did not single out for special appreciation Academic Press director, Dr Conrad Guettler, for his professionalism, ability to get the job done, willingness to work effectively with ISWA, and finally his many personal kindnesses, all of which have enabled us to produce the very best book that its content allows.

William S. Forester
Chicago, 1987

Table of Contents

ISWA Working Group on Hazardous Waste   v

Preface   vii

CHAPTER 1. An Overview of International Perspectives in Hazardous Waste Management
JOHN H. SKINNER   1

CHAPTER 2. Summary and Analysis of Hazardous Waste Management in ISWA Countries
DAVID C. WILSON and WILLIAM S. FORESTER   11

CHAPTER 3. Hazardous Waste Management in Austria
GERHARD VOGEL and CHRISTOPH SCHARFF   97

CHAPTER 4. Hazardous Waste Management in Denmark
KLAUS MULLER   109

CHAPTER 5. Hazardous Waste Management in the Federal Republic of Germany
GERHARD SIERIG   123

CHAPTER 6. Hazardous Waste Management in France
JEAN-BERNARD LEROY   135

CHAPTER 7. Hazardous Waste Management in Italy
ALBERTO PIEPOLI   153

CHAPTER 8. Hazardous Waste Management in Japan
SACHITO NAITO   161

CHAPTER 9. Hazardous Waste Management in The Netherlands
D. den OUDEN   179

CHAPTER 10. Hazardous Waste Management in Southern Africa
A. C. de BRUIN   189

CHAPTER 11.	Hazardous Waste Management in Spain and Portugal JULIAN URIARTE JAUREGUIXAR	207
CHAPTER 12.	Hazardous Waste Management in Sweden KENNETH ANDERSSON	213
CHAPTER 13.	Hazardous Waste Management in the United Kingdom DAVID C. WILSON	231
CHAPTER 14.	Hazardous Waste Management in the United States of America WILLIAM S. FORESTER	269

1

An Overview of International Perspectives in Hazardous Waste Management

JOHN H. SKINNER

US Environmental Protection Agency, Washington DC, USA

INTRODUCTION

The improvement of hazardous waste management practices is one of the most important environmental issues throughout the world. This is true not only in industrialized countries but, to a greater extent than sometimes realized, in developing countries as well. There is a vast amount of experience that can be shared among countries to improve management practices. The International Solid Waste and Public Cleansing Association (ISWA) recognized this potential and at its 1984 Congress proposed a program of primary emphasis in this area (1).

Following the Congress, an ISWA Working Group on Hazardous Wastes was established with representatives from 12 countries or groups of countries spanning four continents. The working group decided that as a first task it would be useful to explore in depth the status of hazardous waste management in the represented countries and prepare a report comparing and contrasting various practices.

The members of the working group developed a set of 12 individual

---

(1) W. Schenkel and J. H. Skinner (1985). Hazardous Waste Management Opportunities for ISWA Activities 1984–1988. *Waste Management and Research*, 3, 1–8.

country reports, prepared in a common format and addressed to a common set of issues. A summary report, which synthesizes the information from individual country reports and draws out common themes, was developed. Also prepared was a special section dealing with hazardous waste management in developing countries. The entire document represents 2 years of effort by the working group.

In developing this document, the working group sought to provide a critical assessment of how national waste management systems are working in practice. Emphasis was not only given to the successes of the national control system but to the remaining problems as well. We hope that this sharing of experiences makes a contribution towards improving hazardous waste management worldwide and that other countries will benefit from our discussions of successes and failures.

The major findings and conclusions of the working group report can be organised into four sections, covering the following major topics:

- *Regulatory Control Programs*, dealing with legislative and regulatory approaches to the control of hazardous wastes.

- *Old and Abandoned Facilities*, dealing with approaches for controlling releases of hazardous wastes from old and abandoned facilities.

- *Treatment, Disposal and Recycling Facilities*, concerning approaches for establishing an adequate number and capacity of facilities to treat, dispose or recycle hazardous wastes.

- *Future Needs and Directions*, concerning uncontrolled waste management practices and future trends and directions of various national programs.

NATIONAL CONTROL PROGRAMS

The establishment of a national regulatory control program with appropriate legislation, regulations, ordinances and licenses is the singular most important step in protecting human health and the environment from the mismanagement of hazardous wastes.

In those countries where regulatory programs are established, clear improvement in waste management practices has occurred. In those countries without regulatory programs, hazardous wastes tend to be dumped indiscriminately or disposed of with household wastes. Many instances of environmental damage and citizen exposure have been cited. Furthermore, in the absence of regulatory controls, adequate incineration and other treatment facilities are not developed. Several countries without regulatory programs expressed concern about becoming a dumping ground for imported wastes.

Several good examples of effective national regulatory programs are featured in the working group reports. While they may differ in their design features, most share the following common elements:

- *Definitions of Hazardous Wastes*, identifying those wastes subject to regulatory control. In some countries, such as the UK and Japan, the definitions are very broad, covering all industrial wastes and even household wastes. In others, such as the USA, the definition is more narrow and the federal control system only applies to specified hazardous wastes.

- *A Registration or Notification Program*, whereby generators, transporters and operators of hazardous waste facilities must be registered with the control authorities. This registry of regulated entities is an important step in establishing a national data management and information system. It also is essential for managing inspection, licensing and enforcement activities.

- *Responsibilities of Waste Generators or Producers*, including responsibility for identifying hazardous wastes, assuring proper packaging and labeling for transport, and tracking wastes to their points of final disposition. Some countries go beyond this. In The Netherlands, for example, each generator is responsible for notifying the regulatory authorities about every transfer of hazardous waste.

- *Responsibilities for Waste Transporters*, including not only requirements for safe waste transport but also a manifest or trip ticket system. The manifest should accompany all waste transits and is the paperwork record for identifying lost or misdirected shipments. Several countries, such as the Federal Republic of Germany and Austria, report very good success with their trip ticket system.

- *Permits or Licenses for Facilities*, were found to be indispensable in most national control programs, each facility that treats, stores or disposes of hazardous waste requires a permit or license from the regulatory authority. The license grants permission to operate the facility and imposes design, operating and maintenance conditions.

- *Control of Waste Import and Export*, in that controls are necessary to assure that imported wastes enter the national regulatory system and that receiving countries are prepared to deal with exported wastes.

In most countries, the responsibility for managing the national control system is shared among national, regional and local governments. While the details differ from country to country, the national government is generally responsible for establishing national standards, guidelines or codes of practice. Regional and local governments are often responsible for enforcement and licensing activities. In some countries, such as the USA, the national government has responsibility to oversee regional government actions and may take independent enforcement or permitting actions as well.

The report for Sweden points out that, even with extensive legislation, certain companies with hazardous waste deliberately or inadvertently bypass the law. The report states that greater resources must be provided for information and supervision, otherwise conscientious companies do not work under the same conditions as the less conscientious ones.

In the UK where enforcement is the responsibility of 160 local waste disposal authorities it was found that existing controls were applied very unevenly across the country. The standards for site licensing, inspection and enforcement were found to vary widely among authorities. There were distortions in the market place, a downward pressure on prices and a leveling down in standards. To remedy this situation, a national Hazardous Waste Inspectorate was established to work towards improvement in the standards for site licensing and closing of loopholes.

This points to the need for frequent and comprehensive inspections and appropriate response to violations including fines, penalties and criminal prosecution, if necessary. Furthermore, technical guidance, technical assistance and training are important for both disposal facility operators and regulatory authorities.

OLD AND ABANDONED FACILITIES

Several countries have had experience in successfully responding to releases of hazardous materials from old and abandoned facilities. Although the source of contamination in Seveso, Italy, in 1985, was a reactor explosion rather than an abandoned facility, the nature and extent of contamination were similar and the technical remedy provides a good illustration. Large amounts of soils and debris contaminated with dioxins were disposed of in two basins lined with HDPE geomembranes, bentonite and clay. The basins were underlain with an inspection tunnel, underdrains and monitoring wells. They were covered with reinforced concrete tops and a soil and vegetation layer. Today, the surface of the accident area is a recreational park, new homes have been built nearby and the situation has returned to normal. This is evidence that successful site remediation can take place and will be accepted by the public.

On the other hand, the residues from the Seveso reactor were shipped out of the country and were temporarily lost in transit. This widely publicized incident points out the need for more effective control of international waste shipments.

In Japan, problems from old hazardous waste sites are not extensive. However, there was one instance where an industrial waste, chromate waste slag, was used as a land reclamation material and resulted in soil pollution by hexavalent chromium. The contaminated solid was successfully cleaned up using a reducing agent, ferrous sulphate.

1. OVERVIEW

While several countries, such as Denmark and The Netherlands, recognize the need to develop programs to deal with old and abandoned sites, the Superfund program in the USA is the only comprehensive national program currently in existence. The Superfund legislation, passed in 1980 and amended in 1986, establishes a $9 billion trust fund to pay for the clean-up of hazardous waste sites. In addition, the government can take legal actions to force those responsible to clean-up or pay for clean-up of those sites. The trust fund is financed by a series of taxes on chemical facilities and corporations and funds received from private parties.

Under Superfund, emergency removal actions can be taken at any site, but longer-term remedial actions can only be taken at sites identified as national priorities. As of October, 1986, the US Environmental Protection Agency compiled information on more than 25,000 sites where hazardous wastes were deposited or discharged in the past. Preliminary assessments have been made at over 18,000 of these sites and 888 have been placed on a national priority list of sites potentially in need of remedial actions. Since the start of the program in 1980, detailed investigation and planning for remedial work has begun at 473 national priority sites, remedial designs have been completed at 110 sites, and remedial actions have been carried out at 143 sites.

In addition to increased funding, the 1986 amendments also made several important changes to the Superfund law as follows:

- There is increased emphasis on the use of destruction, treatment and resource recovery technologies, rather than containment of wastes or disposal on the land.

- The trust fund includes $500 million to clean-up leaks from underground tanks that store petroleum or hazardous substances.

- The law includes a special set of provisions to deal with the prevention and mitigation of the accidental release of hazardous substances from industrial processes (such as occurred in the tragedy in Bhopal, India, and in Seveso, Italy).

TREATMENT, DISPOSAL AND RECYCLING FACILITIES

The most centralized approach toward providing adequate treatment, disposal and recycling facilities was found in Denmark. The Kommunekemi treatment facility in Nyborg serves the entire country. The Danish municipalities are the shareholders of this publicly owned plant, which is supplied with wastes from 300 collection points (at least one in each municipality) and 21 transfer stations. Generators are obliged to deliver their wastes to the collection points. Then they are transported to the central plant over the Danish rail system. The collection points also accept household hazardous wastes and waste oils. The Kommunekemi facility includes a rotary kiln incinerator, which supplies steam for district heating in Nyborg, and inorganic waste treatment facility; a waste oil recovery system, and a controlled landfill for residues. This successful national system has been operating for more than 20 years and processes in excess of 90,000 tonnes per year of hazardous waste.

In Sweden, the SAKAB plant at Norrtorp serves as its central plant for the treatment of hazardous wastes. This plant is 96% state-owned. It processes approximately 60,000 tonnes per year, or 20% of the country's wastes. The facility includes a rotary kiln incinerator, which burns waste oils, solvents, paints, pesticides, and wastes containing PCBs. Energy recovery provides district heating to the nearby town. There is no national organization for the collection of hazardous wastes in Sweden; SAKAB, however, has regional reception centers in different part of the country.

In The Netherlands, there is a good mix of facilities for treating hazardous waste with more than half the waste incinerated, dewatered, and detoxified. Less than 5% of the waste is disposed in landfills. About 40 permit holders are allowed to handle chemical waste, but most of these are small firms with a specialized market. There is a large central facility, the AVR-Chemie, which is jointly owned by the central government, the City of Rotterdam and eight multinationals. This includes two rotary kiln incinerators with a capacity of 70,000 tonnes per year and a controlled landfill for residues.

In the Federal Republic of Germany, large facilities for the treatment and incineration of hazardous waste on site have been constructed by the major chemical producers. In addition, central treatment plants under local government responsibility are successfully serving the small and medium size companies. More than 120 facilities exist in the FRG. These include 27 incineration facilities, with a combined capacity of 620,000 tonnes per year; 23 physical/chemical treatment plants, and 22 special waste landfills.

In the FRG, responsibility for implementing the hazardous waste management program lies with state (regional) governments. In Bavaria, special collection points have been established in order to reduce transport distances for generators and allow collection of sufficient waste for economic transport. In the state of Hessen, a unique facility, Herfa Neurode, is located in an abandoned salt mine. About 45,000 tonnes of waste are deposited in this facility each year, and the storage in the mine is recorded and compartmentalized so that the wastes can be moved later for recycling.

The country report for France states that more than 80 per cent of the toxic and dangerous wastes generated are treated or disposed of properly. In current operation are a diversified mix of incinerators, physical/chemical treatment facilities, oil and solvent treatment centers and controlled landfills. Six river authorities in France are in charge of control of liquid discharges of hazardous wastes. A subsidy system has been established to encourage the treatment of toxic wastes where such treatment is more expensive than landfill.

Funds are apportioned by the river authorities, and subsidies range from 30% to 60% of treatment costs. The river authorities must approve the treatment of the wastes. In addition, a national agency, ANRED (National Agency for Waste Recovery and Removal), has been established to promote new treatment and recovery techniques and centralize information. The river authorities and ANRED, through loans, subsidies and research and development, encourage improvements in methods for treating and recycling hazardous wastes.

In the UK 85% of the hazardous waste is disposed of in landfills. There are no rotary kiln incinerators and there is a shortfall in incineration capacity for capacitors and transformers containing PCBs. The common option is for certain hazardous wastes to be codisposed with domestic refuse in sanitary landfills. While codisposal occurs in other countries due to loopholes or violations of regulatory controls, in the UK it is an authorized practice for certain wastes. There is little evidence of problem landfill sites in the UK on the scale reported in some other countries. This is attributed to comprehensive land use controls, favorable geology and control of ground water usage.

Disposal of waste in or onto the land is also quite common in the USA. Two technologies that are used extensively in the USA, but in very few other countries, are deep well injection and surface impoundments. More than 30 million tonnes per year of liquid hazardous wastes are injected into 90 deep wells. In addition 130 million tonnes per year are treated, stored or disposed of in approximately 200 landfills.

Recognizing the potential for all of these practices to release hazardous substances to surface and ground waters, the disposal of hazardous waste into or on the land is being phased out in the USA. Recent legislation states that the disposal of hazardous waste is prohibited unless such a prohibition is not required in order to protect human health and the environment. It must be positively determined that land disposal of hazardous waste can be accomplished safely. Otherwise, that disposal is prohibited by national law. This should result in an increase in incineration, and physical, chemical, and biological treatment of wastes combined with stabilization of residues prior to the disposal.

FUTURE NEEDS AND DIRECTIONS

The working group countries identified several critically important issues that must receive attention in the near future:

- The development and implementation of national regulatory programs, especially in those countries where none exist.

- The establishment of techniques to deal with certain problem wastes.
- The development of programs to deal with household hazardous wastes.
- The encouragement of waste reduction efforts.

The reports dealing with Spain, Portugal, and the Southern African countries indicate the absence of national control systems for managing hazardous wastes and point out the need for future efforts in this area. In Italy, a regulatory program was recently established, but not much progress toward implementation has been achieved. Also in Italy, a network of treatment and disposal plants able to comply with recent standards is needed. The Southern African report stressed the importance of technical assistance and training to improve the quality of waste management practices, especially in developing countries. The Institute of Waste Management (Southern Africa) has held a number of seminars and technology transfer workshops and the Council for Scientific and Industrial Research has developed research projects designed to solve waste management problems. More efforts along these lines are needed.

In those countries with national regulatory control programs, one common shortcoming was the lack of adequate data collection and information management systems. As a result, statistics on waste generation, treatment, and disposal quantities are very unreliable. Information on permit and compliance status was also very poor. Significant improvement is needed in virtually every country.

The working group members were asked to identify problem wastes for which treatment or disposal technologies were not adequate or available. The responses indicated additional research and technology development is necessary, and that opportunities exist for technology transfer and exchange programs among countries.

While several countries provide collection centers for household hazardous wastes, it seems that a much more concerted effort is necessary to remove hazardous substances from the household waste stream. Two successful programs, in Austria and Japan, appear in those respective reports. In Vienna, mobile hazardous waste collection units were employed and a 50% reduction of hazardous components in the waste was achieved.

In Japan, concern about the emissions of mercury from municipal waste incinerators has led to separate collection of mercury batteries in various municipalities. The national government has recommended strongly that municipalities increase the recovery rate of batteries. They also recommended that battery manufacturers reduce the content of mercury in alkali batteries and develop and popularize batteries using alternative materials.

Many of the working group reports recognized the importance of waste reduction activities by the waste producer. Waste reduction is different from recycling in that it involves reduction or elimination of wastes within an industrial production process. This can be accomplished by process modifications, feed-stock substitutions, and various management practices to increase the efficiency of production. Research studies and information programs are discussed in several of the reports. A few isolated examples are discussed, such as the move from solvent-based paints to water-based paints in Sweden. Conditions attached to permits in The Netherlands and the USA attempt to encourage reduction of wastes. In Denmark, there are demonstration and pilot projects in the field of low- and nonwaste technology with financing up to 100% of costs. However, this is certainly one area that could receive far more attention. Perhaps the most important need for all national hazardous waste control programs is to take this noble concept, which everyone seems to appreciate, and put it into practice.

2

Summary and Analysis of Hazardous Waste Management in ISWA Countries

DAVID C. WILSON

Environmental Resources Limited, London, UK

and

WILLIAM S. FORESTER[*]

American Public Works Association, Chicago, USA

INTRODUCTION

Background

Under the auspices of the ISWA Scientific and Technical Committee (STC), a Working Group on Hazardous Wastes was set up, and met for the first time in Milan in March 1985. By the end of 1986, the working group had held five meetings of 1–2 days each. The following summary report was compiled from country reports prepared by the 12 members of the working group.

During recent years, several international organizations have taken an interest in hazardous waste management. Some of their activities are

---

[*] Author of DEVELOPING COUNTRIES section only. Copyright © 1987 United States Government.

summarized in Table 1. In reviewing these efforts, the working group made a number of observations.

A particular focus in recent years has been on developing controls for transfrontier shipments of hazardous waste. Associated with this has been extensive efforts to harmonize classifications and nomenclatures for hazardous waste in different countries. Also, a number of studies have focused on policy guidelines and on topics to be included in national legislation.

A specific field of international activity has been information exchange in low- and nonwaste technologies. In addition, there have been a number of technical studies that have examined in depth such matters as alternative treatment technologies or research on landfilling in different countries. Finally, a good comparative review of legislative requirements has been produced for the OECD countries.

A major difference between ISWA and the various governmental organizations listed in Table 1 is that ISWA is an organization of professionals, representing practitioners as well as government officials and academics. It is, thus, in a unique position to look below the surface, examining national control systems for hazardous waste management not only as they exist on paper but as they actually operate in practice.

The first task the working group set for itself was to examine in depth the status of hazardous waste management in the countries, or groups of countries, they represent. The main purpose was to share experiences, enabling individual countries to learn not only from their own successes and failures but also from those of others.

Each member produced a country report, following a common structure agreed upon at the first working group meeting. These draft reports were then reviewed at the following two meetings and subsequently revised.

At the third working group meeting, a round-table discussion was initiated by each member summarizing for his own country: things done well; things not done well; and the remaining problems.

A refreshing feature of both the country reports and this discussion was the readiness of members to discuss failures as well as successes and to provide some insight into why things did not work.

From this discussion, about 40 issues of concern, which can be combined under some seven or eight general themes, emerged. This working group report is a synthesis of those themes that emerged from the country reports and from the working group sessions.

Outline of this report

The working group report comprises three main parts: an introduction; the summary report, synthesizing the information from the country reports and drawing out common themes and major issues; and the individual country reports.

TABLE 1
Summary of Activity by International Organizations in Hazardous Waste Management

Organization	Dates	Activities
World Health Organization (WHO) and United Nations Environment Programme (UNEP)	1980–83	Policy guidelines and code of practice (1) — sets out principles for formulating and implementing a policy.
UNEP	1980– 1984–	Transfrontier shipment with regard to developing countries. Ad hoc Working Group on Environmentally Sound Management — developing high level guidance on policies and legislation ("States should ensure.../promote.../take such steps...").
UNEP/World Bank/WHO	1985–	Technical manual for developing countries.
UN Economic Commission for Europe (ECE)	1980– 1985– 1986–	Working Party on Low- and Non-waste Technology and Reutilization and Recycling of Waste — compendium of technologies — analysis of incentives and policy measures — recovery and reutilization of hazardous waste — possible extension of activities to include new technologies for treatment and disposal of hazardous waste and applications of risk assessment.
Organization for Economic Cooperation and Development (OECD)	1974– 1980–83 1984–	Waste Management Policy Group Hazardous waste — abandoned sites — economics — comparative analysis of regulations (2) Transfrontier movements of hazardous waste, developing a binding agreement. Related work includes harmonization of nomenclature, development of an agreed list of hazardous waste and studying the interface between hazardous waste and materials destined for recycling.

TABLE 1 (continued)

Organization	Dates	Activities
	1984–	Other topics include: — assessment of sea disposal — management of small quantities of hazardous waste — PCBs — dioxin from waste incineration — policies related to abandoned sites.
Commission of the European Communities (CEC)	1978–	Implementation of the 1978 Directive on Toxic and Dangerous Waste in Member States.
	1979–	Policies to encourage clean technologies and recycling of industrial and hazardous waste.
	1980–	Improving information.
	1984–	Definitions and classifications.
Council for Mutual Economic Assistance (CMEA)		Development of low- and nonwaste technologies.
North Atlantic Treaty Organization, Committee on Challenges to Modern Society (NATO/CCMS)	1973–81	Study on hazardous waste disposal (3). Focused on landfill research.

(1) M.J. Suess and J. W. Huismans (eds), *Management of Hazardous Waste*, WHO Regional Publications, European Series No. 14, Copenhagen, 1983.
(2) J. Hannequart, *Identification of Responsibilities in Hazardous Waste Management*, OECD, Paris, 1985.
(3) J. P. Lehman (ed), *Hazardous Waste Disposal*, Plenum Press, New York, 1982.

2. SUMMARY AND ANALYSIS

Within the summary report, a number of comparative tables are produced. These have a twofold purpose: to summarize information, such as that on the provision of controls in the individual countries; and to guide the readers to the country reports, which provide more detailed information.

By their nature, the tables provide a simplified overview of the situation. They should be regarded primarily as cross-references to the country reports rather than as conveyors of information themselves.

The remaining sections of the summary report are outlined briefly below:

- National Control (Regulatory) Systems provides an overview of national control systems, examining both the elements within a system and alternative national strategies for implementation.

- Definitions and Quantities of Hazardous Wastes compiles information on quantities produced and examines national systems for data management.

- Collection and Transport focuses on the import and export of hazardous wastes and a number of special topics including household hazardous wastes and waste oils.

- Storage, Treatment and Disposal compares facilities in different countries. Policies are examined for the siting of new facilities and for the encouragement of waste avoidance or recycling. In addition, national programs for the clean-up of abandoned hazardous waste sites are examined.

- Hazardous Waste Management in Developing Countries examines specific problems in those countries.

- Assessment considers accomplishments and problems of national control systems. The major problems and issues to emerge from the working group's deliberations are discussed in turn, and trends for the future are suggested.

NATIONAL CONTROL (REGULATORY) SYSTEMS

Introduction

The terms "control system" and "regulatory system" are for the most part interchangeable. During its period of report development, the working group adopted "control system", and this usage is continued in the published report. Any national control system for hazardous waste must provide three vital components if it is to be successful: legislation introducing a legal framework; proper implementation and enforcement; and adequate facilities with measures to encourage their use.

All three components are vital to the proper working of a national control system. No matter how perfect a system may appear on paper, it is worthless if it is not enforced. Similarly, control cannot be enforced if adequate facilities, for example for high temperature incineration and chemical treatment, are not available.

There is no single control system for hazardous waste that will work perfectly in all countries. The legal, political, and cultural system in each country demands a unique national solution. In this report, an attempt is made to highlight particular characteristics of national systems, highlighting both successes and problems. Features common to a number of systems are identified while concepts unique to particular countries that may be of potential application elsewhere are highlighted.

In the remainder of this chapter, a number of aspects are examined. Among these are the various elements that need to be addressed by an effective national control system for hazardous waste; alternative national strategies for implementing such control systems, including the provision of necessary facilities; and problems of enforcement.

Elements in a national control system

In this section, six of the major elements in a national control system for hazardous waste management are discussed. These include responsibilities placed on the waste generator and registration or licensing of those involved in collection, transport, intermediate storage, treatment or disposal. Other elements are control over transport; permitting of treatment or disposal facilities; planning and establishment of facilities; and programmes for dealing with old or abandoned sites.

The list given above identifies just some of the elements that need to be included within national legislation and control systems. A number of other important regulatory elements are discussed in subsequent sections. These include limitations on the choice of treatment or disposal options for particular types of waste, and approaches to providing adequate facilities and ensuring that they are used. Subsequent discussion items are enforcement, definitions of hazardous wastes, and a national data management system, to ensure that adequate information is available.

In the remainder of this section, each of the six elements listed above are examined for what they contribute to the overall control system, and some of the factors needed to be considered in deciding whether or not to include a particular element in the system are summarized.

Table 2 provides a simplified comparison between the 2 countries or groups of countries represented in the working group. This has a twofold purpose: to put some "flesh" on the board generalizations made in the text; and to serve as a cross-reference and index to topics discussed in the individual country reports. Not all the elements discussed in the text are amenable to simple tabular comparison, so the table is not comprehensive.

TABLE 2
Elements in National Control Systems (1)

	Austria	Denmark	FRG	France	Italy	Japan	Netherlands	Spain	Sweden	Southern Africa	UK	USA
State of progress												
Date of main legislation	1983	1972	1972	1975	1982/84	1970	1979	no system yet enacted	1975	proposed (2)	proposed 1972/74	1976/84
Registration/licensing (3)												
Collectors/transporters	L	(4)	L	R	L	L	No	—	L	No	No	R
Treatment/disposal contractors	L	L	L	R	No	L	L	—	L	No	No	R
Control over transport												
Manifest system	Yes	Yes	Yes	New	Yes	No	Yes	—	Soon	No	Yes	Yes
Control over import	Yes	Yes	Yes	Yes	Yes	Yes	Yes	—	Yes	by sea	Soon	Yes
Control over export	No	Yes	Yes	Soon	Yes	Yes	Yes	—	Yes	by sea	Soon	Yes
Permitting of facilities												
Storage	Yes(5)	Yes	Yes	Yes	Yes	Yes	Yes	—	Yes	Soon	Soon	Yes
Treatment	Yes	Yes	Yes	Yes	Yes	Yes	Yes	—	Yes	Soon	Yes	Yes
Disposal	Yes	Yes	Yes	Yes	Yes	Yes	Yes	—	Yes	Soon	Yes	Yes
Have all operating sites now been permitted	No	Yes	Yes	Yes	No	Yes	Yes	—	Yes	No	Yes	No

TABLE 2 (continued)

	Austria	Denmark	FRG	France	Italy	Japan	Netherlands	Spain	Sweden	Southern Africa	UK	USA
Planning and establishment of facilities												
Is there a national strategy/plan?	Yes	Yes	No	No	No	No	Yes	—	No	No	No	No
Are authorities required to produce a plan?	Yes	Yes	Yes	No	Yes	Yes	Yes	—	No	Yes	Yes	No
Has this been done?	Yes	Yes	Yes	No	No	Partial	Yes	—	No	No	Partial	No
Old or abandoned hazardous waste sites												
Is there a national inventory	New	Yes	Yes	Yes	No	No	Yes	—	Yes	No	No	Yes
Is there a clean-up programme?	No	Yes	(6)	(6)	No	No	Yes	—	Yes	No	No	Yes

(1) For explanations, see text.
(2) In Southern Africa, the situation varies between countries. In most cases, control is presently informal, in the absence of formal legislation.
(3) L = licensing scheme, implying investigation by the authorities; R = registration, implying simply being listed in a register.
(4) Partial, from the central collection points to the treatment plant.
(5) Mainly under the Trade Act, not under the Hazardous Waste Act.
(6) Although there is no formal, nationwide clean-up programme, the clean-up of individual sites is proceeding.

2. SUMMARY AND ANALYSIS

One additional item included in the table is the date of the main legislation of hazardous waste. This is significant, as implementation of a control system normally takes several years. Thus, in Austria and Italy for example, it is too early to form judgements on how the controls are working in practice.

Responsibility placed on the waste generator

In most countries, the responsibilities of the hazardous waste generator are defined in the legislation (1). The exact definition of these responsibilities varies from country to country, but elements include the following:

- to know the waste and decide whether it comes within the control system;
- to declare production of hazardous waste to the authorities;
- to nominate an individual as hazardous waste liaison officer;
- to prepare the manifest or consignment note to accompany each load of waste;
- to pack and label the waste in accordance with regulations on transport of hazardous materials;
- to keep records;
- to ensure that the waste is delivered to a treatment or disposal facility licensed to accept it; and
- to make regular reports to the authorities on quantities, composition and treatment/disposal methods.

In most countries, the waste generator has fulfilled his responsibility if he has accurately disclosed the composition of his waste and delivered it to a treatment or disposal facility licensed to receive it. The situation when the waste is passed to a transport contractor is sometimes less clear cut.

Exceptions include the USA where fault liability applies. Thus, if a properly approved facility was to become a problem site in the future, then producers whose waste had been delivered so the site would be liable for clean-up costs. The position is similar in Denmark for those wastes not delivered to the national treatment facility.

Registration or licensing of parties

A possible element in the national control system is a registration or licensing system for waste collectors and transporters, and/or for waste treatment and disposal contractors.

---

(1) The differences between standards and controls applied to prime products and those applied to wastes are beyond the scope of this report, which focuses solely on wastes.

Registration is a relatively simple procedure, whereby contractors are required to register with the authorities. No endorsement of their competence is required on the part of the authorities, although in some countries any contravention of legal requirements may result in the firm being struck from the register.

Licensing is a more complex procedure, requiring the authorities to examine the professional, technical and financial competence of the contractor to carry out the duties for which he requires a license.

As shown in Table 2, registration of operators is more common than licensing, although neither is universal. Arguments against include: the additional burden on controlling authorities for administering a registration, and more particularly a licensing scheme. A second consideration includes the duplication of effort, given that the credentials of the operator will generally be one of the criteria used in assessing applications for permits to run treatment and disposal facilities.

Control over transport

Hazardous wastes often have to be transported long distances to a treatment or disposal facility. It follows that there could be a strong financial incentive for an unscrupulous operator to dump the waste illegally, either at a local landfill site not licensed to receive hazardous waste or even by the roadside, or to mix the hazardous waste with other wastes prior to local landfill or incineration.

In order to prevent such abuses, most countries have or will soon implement a system for cradle-to-grave control, through some kind of manifest system. The individual systems in different countries differ in the detail of their operation and are discussed more fully in a later section.

International attention has been drawn to the potential problems of import and export of hazardous wastes by two events:

- the export in 1982 from Italy of 41 drums of Seveso dioxin waste, which were eventually discovered in a warehouse in France; and

- reported proposals to export hazardous wastes from developed countries for disposal in developing countries.

As a result, a number of international organizations, notably the OECD, UNEP, and the Commission of the European Communities, are working towards international agreements on manifest systems to control transfrontier shipment of waste. As shown in Table 2 a number of countries already, or will soon, include control over exports and imports in their national control systems.

2. SUMMARY AND ANALYSIS

Planning, permitting and old site clean-ups

A basic requirement of any control system is that facilities for waste treatment and disposal should be licensed or permitted by the authority. All the national control systems surveyed incorporate a requirement of this kind.

The extension of this permitting requirement to storage facilities has, until recently, been less general. Some exemption is usual for temporary storage on a producer's premises pending collection, but care is required to ensure that such exemptions do not allow storage under insecure conditions or the accumulation of large quantities of waste pending economic recovery at some unspecified date in the future.

Table 2 includes a question as to whether all existing facilities have been permitted under the (new) regulations. This provides an indication of the state of implementation of the system.

Another important feature of national control systems is a measure to ensure that adequate facilities are available both now and in the future for the treatment and disposal of all hazardous waste. The various national approaches to achieve this objective are discussed in the next section.

It can be argued that, whatever method is used, it is important that coherent plans should be produced, taking into account the types of quantities of waste expected over the next, say, 20 years and the facilities that will be provided to deal with them.

These plans need to be prepared at both the national and regional or local level. Preparation of such plans is required in legislation in a number of countries. As noted in Table 2, however, even where this is the case progress in actually producing the plans has often been poor.

Discussion so far has centered on the control of hazardous wastes currently being produced. A complementary aspect of national control systems in a number of countries is the investigation and clean-up of old sites containing hazardous waste that may pose a threat to public health, either through direct contact or through the contamination of groundwater resources. The priority given to this element in the overall hazardous waste programme depends on national circumstances and the extent of perceived problems.

Different national approaches to implementing a control system

Despite some measure of agreement on the significant elements to be incorporated in a national control system, there are probably as many different approaches to implementing such a system as there are nations that have done so. The purpose of this section is to highlight some of these differences in philosophy and approach, pointing out both success stories and problems encountered.

The discussion is organized around three particular features of a national system:

- the extent to which control is devolved, or delegated, from national to regional and local governments;
- the methods used to restrict the choice of treatment and disposal options for particular types of waste; and
- the approach to the provision and financial support of treatment and disposal facilities.

Devolution of control

In most countries, primary responsibility for the control of hazardous waste is devolved either to regional or local government. However, the extent of this devolution varies considerably, both in the numbers and size of the controlling authorities and in the degree of national control exercised over them.

In the USA the national government is directed by law to delegate control to state governments, if the state can demonstrate its capacity to undertake the function. Rigid national standards are set down for permitting facilities, and state programs may be no less stringent.

In the UK, control is devolved to nearly 200 local authorities, some of them being extremely small, with little or no technical expertise. Control is exercised primarily through site licensing or permitting, the conditions of which are at the discretion of the local authorities. The role of national government is largely advisory, although they have published an extensive series of Codes and Practice for the treatment and disposal of different types of hazardous waste, each one being based on the report of a working group comprising representatives of government, industry, controlling authorities, and other interested parties.

The situation in the Federal Republic of Germany (FRG), Austria and Japan is somewhat intermediate, with primary responsibility being again at the regional (state, provincial or perfectural) level, but with the regions having rather more discretion than in the USA. In these cases, national governments provide guidelines, including technical norms for the avoidance, treatment, or disposal of, particular categories of waste.

Choice of treatment or disposal options

The various national systems differ in the degree and methods of control over the choice of treatment or disposal option for a specific waste. Three broad alternatives for exercising such control may be distinguished.

2. SUMMARY AND ANALYSIS

First, central (or regional) government may assume powers to direct a specific type of waste either to a particular treatment or disposal option or even to a particular site.

This is effectively the case where the waste is passed to a state utility (as in Denmark and generally also in Sweden). Otherwise there is a possible problem, as the responsibility for deciding the most appropriate option for a particular waste is shifted from the producer to the state.

In Japan, the treatment or disposal route to be used for certain waste types is specified in regulations. In Austria and FRG, recommendations are made in technical standards, but these are not mandatory.

Some countries, the UK and Japan, have included a power of direction to a specific site in legislation, to be held in reserve for use should the need arise.

Secondly, an alternative to direction is to prohibit the use of certain options for particular wastes. Such prohibitions are used in most of the countries on the working group, particularly regarding land disposal. Examples include the USA and The Netherlands. In the latter case all land disposal of hazardous waste is prohibited unless specific exemption is given.

The third option, which is used by all countries to a certain extent, is control via site permits. Waste can only go to a facility permitted to receive it, so that if, for example, the permits for all landfill sites exclude a particular category of waste, then that waste has effectively been prohibited from land disposal. However, variations in standards of site permitting between local controlling authorities often mitigates against the effectiveness of such control on a national basis. This is particularly so in the UK, but applies also in a number of other countries, including France and Austria.

A comparison of methods for regulating the choice of treatment and disposal options in the various countries is shown in Table 3.

Provision of treatment and disposal facilities

As discussed above, a necessary part of a national control system is to provide adequate facilities for the treatment and disposal of hazardous waste and to ensure that the proper facilities are used.

This problem relates primarily to "high technology" facilities for incineration and chemical treatment, the costs for which are often an order of magnitude greater than those for landfill disposal. If landfill and treatment facilities are allowed to compete for the same wastes, then treatment facilities are unlikely to be either financially viable or attractive to the private investor.

When faced with a stubborn problem, the appropriate solution is often similar to that traditionally used to encourage the stubborn donkey to move forward: that is, a combination of a carrot in front and a stick at the rear. In this context, the "stick" is provided by restricting the choice of

TABLE 3
Regulating the Choice of Treatment or Disposal Options for a Specific Waste (1)

	Austria	Denmark	FRG	France	Italy	Japan	Netherlands	Spain	Sweden	Southern Africa	UK	USA
Direction of waste												
To (a) particular site(s)	No	Yes	No	No	No	No	No	no system at present	Yes	No	No	No
To (a) particular option(s)	(2)	Yes	(2)	No	No	Yes	No		Yes	No	No	No
Powers exist, in reserve	No	—	No	Yes	No	Yes	No		—	No	Yes	No
Prohibition of certain options for particular wastes												
National regulations	(2)	Yes	Yes	Yes	Yes	Yes	Yes		Yes	No	No	Yes
Control via site permits												
Strong national standards mean effective prohibition for certain waste	Yes/No (3)	Yes	Yes/No (3)	Yes	(4)	Yes/No (3)	Yes		Yes	No	No	Yes

(1) See discussion in text.
(2) Recommendations are made in technical norms, but these are not mandatory.
(3) Strong controls exist in principle, but in practice there are wide local variations in what is/is not permitted at individual sites.
(4) System not yet in place.

alternatives for any particular waste, while the "carrot" is provided by some form of financial support of the high technology facility.

Looking first at the provision of facilities, a number of cases may be distinguished. In Denmark and Sweden, major treatment facilities are provided by a state utility, which is a co-operative venture between central government, local communities and industry.

In a number of other countries, the public and private sectors also collaborate in providing facilities, often with substantial investment from government (e.g. the Netherlands). In France, the facilities are entirely provided by the private sector, but substantial finance has been made available on attractive terms by government. Of the countries with well-established systems, only the UK and USA rely entirely on the private sector for the provision of facilities.

Turning now to pricing policies for treatment facilities, most countries reinforce the "stick" of restricted choice of options with the "carrot" of financial support. The major form of financial support is through indirect subsidy, that is by providing low cost finance for the facility which reduces (or even eliminates) the capital charges which must be met from revenues.

One example of a more direct approach to subsidising the use of the most appropriate or "best practicable" means of treatment or disposal for a particular waste is provided by France. The regional water authorities provide a subsidy of about 30–50%, which is paid directly to the site operator. The necessary revenue is currently raised from charges for waste water treatment, but alternative, more appropriate revenue sources are being studied.

Another form of direct financial incentive for the use of recycling or treatment facilities is to levy a graduated tax on waste production or disposal, so as to provide an incentive for choosing a "better" method. Thus a zero tax is levied on recycling, a maximum tax on landfill and perhaps a 30–50% tax on treatment. A system of this kind has been developed, for example, in the Flanders Region of Belgium. A similar system has been tried in The Netherlands, but may soon be phased out (1).

A few countries do not use the "stick and carrot" model to encourage the use of treatment facilities.

In the USA, reliance is placed mainly on the "stick" (1), through rigid regulations that set down exactly what option may be used for a particular waste. It is then left to the private sector to provide the necessary facilities.

Once the control system is fully established, this mechanism should be effective. However, at present many of the facilities that existed before the regulations were introduced are working on "interim status", due to the slowness in permitting these facilities. This creates a disincentive to invest either in treatment plants or in state-of-the-art landfill sites that satisfy the new regulations, since older, less environmentally sound and therefore cheaper sites, may still exist in neighboring states.

---

(1) At the time of this writing, the US Congress is considering a graduated tax on waste disposal and treatment. Some states already have such a tax.

In the UK, reliance also is placed on market forces to provide treatment and disposal facilities. However, in this case regulatory control over the choice of options is weak. As previously discussed, control is achieved primarily through site licensing or permitting, with the conditions being left largely to the discretion of local authorities. Some authorities take a more lenient view than others as to what wastes may be allowed in landfills. According to the Hazardous Wastes Inspectorate, recently set up to oversee standards across the country, all existing merchant incineration and treatment facilities in the UK are on an economic knife edge as a result of "unfair competition" from landfills.

Table 4 summarizes the provision of treatment facilities and the pricing policies in the various countries.

Enforcement

The third vital element in national control systems, after providing appropriate regulations and adequate facilities, is to implement and enforce the system.

As discussed above, there is a wide diversity of national approaches to the control of hazardous waste, each of which addresses in a slightly different way the central problem of how to encourage compliance with the regulations. Whatever the approach or "compliance strategy" adopted, it appears most countries that have implemented a control system are facing difficulties with enforcement.

In Denmark and Sweden, collection and transportation of hazardous waste are handled through national or local utility companies. In both cases there are problems with the level of supervision of waste producers, in that both staff and expertise are lacking in small communities.

In The Netherlands, enforcement was originally a national responsibility. Inadequate performance, however, has led to that responsibility being delegated to provincial (regional) governments. A 4-year programme, designed to improve observance of regulations and governments' ability to undertake their new responsibilities, has been established.

In the UK, enforcement is undertaken by 165 local authorities, which vary widely in their resource and expertise capabilities. A small, centralized Hazardous Wastes Inspectorate has recently been set up to oversee standards across the country. They have found that existing controls are very unevenly applied, leading to "gross disparities" in the standards of hazardous waste disposal operations in different parts of the country.

In Austria, responsibility is in the hands of nine provinces, but coordination and lack of resources and personnel is still considered to be a major problem.

In the USA, enforcement and compliance with the regulations are considered to be inadequate, with problems in the availability and training

TABLE 4
Provision of Treatment Facilities (1)

	Austria	Denmark	FRG	France	Italy	Japan	Netherlands	Spain	Sweden	Southern Africa	UK	USA
State utility company	No	Yes	In some States	No	No	No	No	No	Yes	No	No	No
Provision of facilities												
Private sector	New	No	Yes	Yes	Yes	Yes	Yes	Yes	Yes	Yes	Yes	Yes
Public sector	No	No	Yes	No	No	Yes	Yes	No	Yes	Yes	No	No
Public/private collaboration	Yes	Yes	Yes	Yes	(2)	Yes	Yes	(3)	Yes	No	No	No
Public investment	Yes	Yes	Yes	Yes	(2)	Yes	Yes	(3)	Yes	No	No	No
Pricing policy												
Indirect subsidy (of investment)	Yes	Yes	Yes	Yes	Yes	Yes	Yes		Yes	No	No	No
Direct subsidy	No	Yes	No	Yes(4)	No	No	No		No	Yes	No	No
Incentive levies/taxes	No	No	No	No	No	No	Yes		No	No	No	some States

(1) For explanations, see discussion in text.
(2) System not yet in place.
(3) No system at present.
(4) France has a system for the direct subsidy of prices charged by a facility, so long as its use is seen as the most appropriate treatment or disposal option for the particular waste in question.

of personnel, in consistency of standards, and in differing priorities for policing in the various sites.

DEFINITIONS AND QUANTITIES OF HAZARDOUS WASTE

Terminology and purpose

Before discussing definitions of hazardous waste, a comment is necessary on nomenclature or terminology. In this report, we use the generic term "hazardous waste", while in various individual countries terms such as chemical, special, poisonous, toxic, or difficult are used to describe such waste.

It is extremely difficult to define a hazardous waste with any precision. Every national system differs both in the detailed method used for definition and also in the breadth of waste included. The importance attached to a legal definition of waste depends both on its purpose and the use that will be made of it.

In many countries, notably the FRG and the USA, one purpose is to separate hazardous from municipal or industrial waste, in order to allow more stringent control. The need for a clear-cut definition is greatest if national policy requires that hazardous wastes be treated and disposed of separately from non-hazardous wastes.

Another common purpose of a definition is to achieve the most appropriate treatment for a particular waste. This, for example, is seen as paramount in Denmark.

In the UK, the primary control over hazardous waste treatment and disposal is through site licensing or permitting, which applies to all household, commercial and industrial waste. Official policy favours the co-disposal of hazardous and municipal waste in landfill sites. Both acceptable types of wastes and the disposal conditions may be included in the terms of the site permit. The only area where additional control over a hazardous waste is felt to be necessary is during transport. The definiton of a "special waste," for which a manifest is required, is thus framed in terms of protecting human health should a waste be illegally dumped in transit, rather than in terms of protecting the environment following proper disposal.

In Japan, the definition of a hazardous waste separates only those wastes for which incineration or chemical treatment is particularly difficult. It is considered that these wastes require total isolation from the environment, by detoxification if appropriate or solidification in concrete and landfilling in a special concrete-lined disposal pit.

International efforts are currently underway to establish a cross-reference list of hazardous wastes as a first step towards harmonization of

2. SUMMARY AND ANALYSIS

definitions. The particular context for this initiative is twofold:

- to ensure the proper working of international controls over transfrontier shipments of hazardous waste; and
- to achieve a distinction between hazardous waste and materials destined for recycling, which ensure proper control over the latter while not creating an unnecessary disincentive to recycling.

Types of definitions

Most countries have used a definition based on three criteria. These are particular types of hazardous wastes, industrial processes from which the waste is defined as hazardous; and/or substances, the presence of which is indicative of potential hazard.

In some cases, a listing on one or more of these criteria is judged to provide a sufficient definiton. In other cases, a further reference is made to particular concentration levels for each of the substances, or to other indicative criteria. These may include:

- the toxicity of the waste itself (as for example in the UK);
- the toxicity of an extract of the waste, usually obtained by means of a specific leaching test. Toxicity is generally defined by reference to concentrations of specific substances in the extract (e.g. in the USA and Japan);
- the ignitability or flammability of the waste;
- the corrosiveness of the waste; and
- the reactivity of the waste.

A simple classification of definitions as used in the various countries is given in Table 5. However, comparison is necessarily inadequate, as it does not give any insight into the breadth of the definitions, the numbers of wastes or processes on the lists, the numbers of substances considered, or the concentration limits.

In the USA and the FRG, several hundred specific hazardous wastes and processes are listed, while the indicative criteria apply to all waste unless specifically excluded.

In Japan, a waste is hazardous only if it is a "cinder, sludge, waste acid, waste alkali, slag or dust" containing one of nine toxic substances above a prescribed limit when tested by an extraction test. In addition, wastes can only be classified as hazardous if they are generated from a specified facility.

In The Netherlands, almost all of the toxic substances have been listed in four categories of concentration levels, that is above 50 mg/kg, 0.5%, 2%

and 5% by weight. Attention is now being paid to waste containing highly toxic substances with concentration levels under 50 mg/kg with a view to deciding whether or not the definition needs to be extended.

One important aspect of a definition is that it should be flexible, so that new types of waste do not evade control. This is a particular problem that must be addressed by any system relying mainly on listing specific types of waste or industrial processes. It is also cited by Denmark as the reason why they do not prescribe concentration limits.

Exclusions

In comparing definitions of hazardous waste, it is important to note those categories of waste specifically excluded from the control system. The reasons for such exclusions are generally concerned with problems of implementation, either because quantities are small or because the political or practical difficulties are too great:

- Hazardous wastes from households are outside the normal control system in all countries.
- Small quantity generators are often placed outside the system. For example, in the USA producers generating less than 1000 kg per month were excluded from control up to 1984, while a new, lower limit of 100 kg per month has now been implemented. It is estimated that this will increase the number of generators covered by the regulations by a factor of 10.
- Aqueous effluents discharged to sewer or treated on-site are controlled separately from hazardous wastes in most countries. However, in the USA waste water treatment in surface impoundments or lagoons is controlled within the hazardous waste regulations.
- Sewage sludge is specifically excluded in some countries but not others.
- Mining wastes are often excluded from control, even though the nature of the waste may be very similar to other hazardous waste.
- Similar exclusions may apply to agricultural waste.
- Certain "hazardous" waste was already controlled in many countries under separate regulations and was therefore excluded from the new hazardous waste controls. Common examples are radioactive, pathogenic, and conventional explosive waste.

Special rules

Attention is drawn here to rules concerning the mixing of hazardous with

TABLE 5
Definitions of Hazardous Waste (1)

	Austria	Denmark	FRG	France	Italy	Japan	Netherlands	Spain	Sweden	Southern Africa	UK	USA
Is there a legal definition?	+(2)	+	+	+	+	+	+	–	+	(3)	+	+
Purpose of definition												
Control over transport	+	+	+	+	+	–	+		+	–	+	+
Control over treatment/disposal	+	+	+	+	+	+	+		+	+	–	+
Type of definition												
List of waste	+	+	+	–	–	+	–		+	–	–	+
List of substances	–	(+)	–	++	++	++	++		++	–	++	++
List of processes	–	–	–	++	++	++	++		–	–	–	++
Concentrations	–	–	–	–	++	–	++		–	–	–	+
Criteria												
– toxicity of waste	+	+	++	++	++	+	–		++	++	++	++
– toxicity of extract	++	++	++	++	++	–	–		++	–	++	++
– ignitability/flammability	–	+	–	+	–	+			–	+	+	+
– corrosiveness	–	++	–	++	–	–			–	++	+	++
– reactivity	–	++	–	++	–	–			–	++	–	++

TABLE 5 (continued)

	Austria	Denmark	FRG	France	Italy	Japan	Netherlands	Spain	Sweden	Southern Africa	UK	USA
Special rules (1)												
Mixing rule	+	+	+	+	+	−	+		−	−	−	+
Residue rule	−	+	−	+	−	−	−		−	−	−	+
Exclusions (1)												
Small generators (1)	−	−	−	100 kg		+	−		−	−	−	100 kg
Waste water	+	−	+	+		+	+		−	+	+	(5)
Sewage sludge	+	−	−	−		+	−		+	+	+	−
Mining waste	+	−	+	+	+	+	−		−	−	+	+(6)
Agricultural waste	+	−	+	+	+	+	−		−	−	+	+

(1) For explanations, see discussion in text.
(2) Key: + = Yes
       − = No
(3) New definition in South Africa only.
(4) Quantity is that per month below which a producer is exempt from the regulations.
(5) Partial exclusion for waste water treated exclusively in permitted treatment tanks. Wastewater treated in surface impoundments or lagoons is controlled as a hazardous waste.
(6) Mining waste excluded pending further study.

non-hazardous waste and also residues resulting from the treatment of hazardous waste.

Where the definition of a hazardous waste depends on the concentration of specific substances, either in the waste itself or in an extract, it is in principle possible to evade control by mixing the hazardous with a non-hazardous waste. This is permitted in some countries but not in others. For example, in the USA, the so-called "mixing rule" states that if you mix a listed hazardous waste with anything else, then the result is a hazardous waste (1). In France, special authorization is required before such mixing, while in The Netherlands, mixing is forbidden.

The treatment of hazardous waste generally results in a residue for final disposal. The question is, whether these residues can be regarded as a non-hazardous waste. Again, in the USA, the so-called "residue rule" states that any residue from treatment of a listed waste (1) is also a hazardous waste unless specifically "delisted", which is a lengthy and onerous process.

In the FRG, there is some confusion regarding the classification of fly ash from municipal incinerators, which some states define as a hazardous waste and others do not; however, the trend is towards its classification as hazardous waste. In other countries, such as The Netherlands, the normal concentration limits or criteria are applied to residues in determining whether or not they are a hazardous waste.

Problems in comparing waste statistics

It is extremely difficult to compare quantities of industrial or hazardous waste in different countries, either accurately or consistently. There are problems with both the definitons used, and in the collection of reliable statistics. The latter problem will be discussed further in the next section. A number of more specific problems may be encountered in comparing particular national estimates.

Particular bulk waste streams may be included in national statistics in some countries and not in others. An example is waste sulfuric acid from titanium dioxide manufacture, which is currently discharged at sea from the majority of manufacturing plants in Europe.

Wastes being discharged at sea may or may not be included. In some cases, sea dumping from barges is included in the statistics, but direct discharge by coastal outfall is generally excluded.

This also illustrates the problem of differentiating between waste and effluents. Effluents are excluded from waste statistics in many countries, but this exclusion is only partial in the USA.

---

(1) The US rule applies to waste listed (by name or process) in the regulations. It does not apply to waste that is hazardous by virtue of the extraction test, which if diluted below the limit is no longer considered as a hazardous waste.

In-house treatment or disposal by the waste producer is not always fully included within national statistics.

Given these comments and the problems of comparing waste statistics between countries, the working group was reluctant to prepare a comparative table, which, despite any caveats applied to it, would almost inevitably be used, as an "authoritative" source of information. However, on reflection it was considered preferable to present the information together with the caveats, rather than simply inviting others to compile the data from the country reports without necessarily realizing its deficiencies.

Table 6 summarizes information on both industrial and hazardous waste quantities in 11 countries represented on the working group. The primary sources of information are the country reports that follow, together with the 1985 OECD Environmental Data Compendium.

The table comprises four columns as follows:

- General industrial waste (**A**), including materials similar to normal municipal or construction waste.

- "Special" industrial waste (**B**), which by its nature needs to be considered separately from "normal" waste. A legal definition of this kind exists in both Austria and France.

- Hazardous waste (**C**), as reported in national statistics.

- Estimates of hazardous waste quantities (**D**), using sources in addition to official figures. For Austria, Italy and Portugal, no official figures are available, so estimated data are reported. For Denmark, the official statistics reflect accurately the quantities of waste known to the system. Column D reports an estimate of total quantity.

Official statistics in the FRG exclude certain bulk waste being dumped at sea. Column D includes an allowance for these. In the UK, the legal definition of "special waste" is restrictive, so column D includes an estimate of "hazardous waste", using the more liberal definition that existed under previous legislation.

Comparing rates of hazardous waste generation

It is instructive to compare the estimates of industrial waste quantites in columns **A** and **B** of Table 6 with those of hazardous waste in column **C** or **D**. The ratio of industrial:hazardous waste is generally in the range of 5:1 to 10:1. Where the radio is either much lower or much higher than this, one suspects that the definition of hazardous waste being used is significantly different from that applied to other countries. From the table, the most obvious discrepancies are for the USA, with a very low ratio, and for Japan, with a very high ratio.

TABLE 6
Comparative Quantities of Industrial and Hazardous Waste (all figures are in million tonnes per year (1))

Country	A General industrial waste	B "Special" industrial waste	C Hazardous waste (national statistics)	D Hazardous waste (see comment column)	Comments
Austria		13	—	(0.1–0.3)	National statistics available for "special" (column D) but not yet for hazardous waste. Estimate in column D taken from OECD data. National statistics available for "special" waste but not yet for hazardous waste.
Denmark	(0.8)		0.12	0.2	91,000 tonnes delivered to the central treatment plant in 1985, 24,000 tonnes exempted. Higher figure in column D is rounded, allowing for an estimated 30% of wastes escaping control.
Federal Republic of Germany	(52)		4.5	(5.5)	Column D includes 1 million tonnes of waste acids dumped at sea.
France	50	18	2.0		

TABLE 6 (continued)

Country	A General industrial waste	B "Special" industrial waste	C Hazardous waste (national statistics)	D Hazardous waste (see comment column)	Comments
Italy	(35)		—	(2–5)	No national statistics available. Column D gives a range based on two alternative estimates.
Japan	292		(0.8)		768,000 tonnes of hazardous waste in 1982. Low figure reflects a rather restrictive definition.
Netherlands	(5)		1.0		
Portugal	(11)		—	2.5	Column D reports estimated data, using waste per employee and waste per unit production data from a survey in France.
Sweden	5		0.5		
UK	(30–40)		2	5	Column C is based on restrictive definition of "special waste", Column D estimates "hazardous waste".
USA	(400)		264		Column C based on 1981 survey.

(1) Data in parentheses does not come from the country reports.
The main external source used is the OECD Environmental Data Compendium 1985 (OECD, Paris).
Column A includes industrial waste similar to normal municipal or construction waste.
Column B includes only "special" industrial (or other) waste, which by its nature needs to be considered separately from "normal" waste. A legal definition of this kind only exists in Austria and France.
Column C is based on official national statistics for hazardous waste.
Column D includes information of hazardous waste, not included in official figures (see comment column).

2. SUMMARY AND ANALYSIS

In the USA the proportion of total industrial waste considered hazardous appears to be high. In addition, the absolute quantity of hazardous waste is much greater than that in other countries, even after allowance has been made for differences in gross domestic product. One explanation is that the statistics for hazardous waste include very large quantities of aqueous effluents, which are either disposed of by deep well injection or are treated in lagoons or surface impoundments prior to discharge to surface waters. Such effluents are generally not included in waste statistics in other countries.

In addition, spills of hazardous chemicals at a number of municipal waste water treatment plants in the USA have, by virtue of the mixing rule, rendered all effluents being treated in those facilities, as well as the resulting sludges, hazardous wastes for the purpose of the statistics.

In Japan, the proportion of general industrial waste considered hazardous is extremely low. As noted previously, the definition of hazardous waste is very restrictive. Most industrial sludges, acids and alkalis, together with all waste oils, are excluded from the statistics.

Comparing waste statistics in more detail

For the purposes of managing hazardous waste, it is essential to have information on the sources and types of waste produced, for example:

- waste breakdown by producing industry:
- waste breakdown by type; and
- waste breakdown by physical form (solid, liquid, sludge).

Unfortunately, national data management systems are such that detailed information of this kind is only available in a minority of cases. General international comparisons are not possible, but the individual country reports do provide some useful information worthy of comment here.

Table 7 summarizes information from four country reports on the major industrial sectors producing hazardous waste. In two cases, the dominant producer is the chemical and petrochemical industry, while in the other two it is the metal-related industries, including metal production, engineering, and electroplating or surface treatment. In all four countries, the percentage of total hazardous waste arising from these two broad industrial groupings is in the range 70-98%.

In the country reports, some information on the catagorization of hazardous waste by type is given in the reports on Sweden, the UK and the USA. Unfortunately, the classification systems used in the three cases are very different, so that direct comparison is not possible. The reports on Japan and Austria include information on the classification of industrial waste by type.

Unfortunately, little information is available on the physical form of hazardous wastes in the country reports. Two exceptions are in the reports on Sweden and the UK from which the data are reproduced below:

	Solid %	Sludge %	Liquid %
Sweden	22	16	62
UK	14	28	58

Wastes escaping control

National statistics on hazardous waste generally focus, by necessity, on that waste reported to the authorities. It is recognized in a number of countries that significant quantities of waste are escaping control and thus are not included in the national statistics.

It is reasonable to suppose that many of these wastes are produced by small quantity generators. As noted above, legislation has recently been amended in the USA to bring producers generating between 100 and 1000 kg per month of hazardous waste within the control system.

In The Netherlands, national statistics include an estimate of about 100,000 tonnes, 10% of the total production of hazardous waste, from small quantity generators, much of it currently escaping control.

In Denmark, the quantity of waste received by the central treatment plant increased by nearly 20% per year between 1983 and 1985. This high figure may indicate an increase in the coverage of the system rather than just an increase in the quantities of hazardous waste produced. It is noted in the country report that the hazardous waste per capita figures are higher in those areas that operate a collection service for hazardous waste, rather than simply relying on the generator to deliver his waste to the collection point. Field trials have suggested that as much as 30% of hazardous waste may currently be escaping control.

In Sweden, some 18% of hazardous waste comes not from industry but from "other activities". As noted in Table 7, this differs significantly from the reported position in other countries. This may be due to real differences in the nature of hazardous waste production in Sweden as compared to other countries, or it may be that the Swedish system is more efficient in collecting hazardous waste from non-industrial activities.

Again in Sweden, data on hazardous waste collected by one municipal monopoly company (in Gothenburg) show a threefold increase from 1981 to 1985, which is attributed both to improvements in the collection service and to additional control efforts.

TABLE 7
Hazardous Waste Production by Industry

All figures as percentages

Industry	France	FRG	Sweden	USA
Chemical/petrochemical	25	70	21	71
Metal related industries (2)	(56)	(28)	(50)	22
Metal production	—	11	22	—
Engineering	16	13	28	—
Electroplating	40	4	—	—
Paint and varnish shops	7	—	—	—
Forestry/agriculture	—	—	2	—
Other industries	} 12	—	19	7
Other activities		—	18	—

(1) Breakdown also given in reports on Japan and Southern Africa for total industrial rather than hazardous wastes.
(2) Figures in parentheses are not used in primary categorization in the country report.

National data management systems

National data management systems may cover information both on waste quantities and on regulated parties such as producers, transporters, and treatment and disposal operators. Table 8 summarizes various national systems for collecting information on waste quantities.

Several countries, including Denmark, the FRG and The Netherlands, have well established national systems that produce information on an annual basis. Information in other countries, such as Sweden and the USA, is based on national surveys carried out some years ago. The UK appears to be alone in having a system for information collection at the local level, but collation at the national level is made difficult.

A number of alternative methods for data collection may be distinguished. Perhaps the most comprehensive is the (annual) report by waste producers, which may form part of a registration scheme. In Denmark, this is viewed, along with trip tickets, as a vital part of the system, providing good information and allowing government to learn more about waste production and so to check the data. A report by treatment and disposal operators will in principle yield the same information, but may not give so much insight as to the origin of the waste.

Where manifests or trip tickets are notified to the authorities, this information could be used as the basis of information on those wastes treated or disposed of outside the factory. A national data management system of this kind is being installed in France.

Sample surveys may be carried out by or on behalf of national governments. These may go directly to producing industries and to treatment and disposal operators, or they may simply seek information from regional or local authorities.

Those regional or local authorities may themselves carry out sample surveys. Such surveys are required in the UK, but no timescale has ever been attached to their completion so progress has been slow. Available results are difficult to collate due to differing classification systems and to wide variation in the dates of the surveys.

COLLECTION AND TRANSPORT

Organization of collection

In most countries, collection and transportation of hazardous waste is carried out by private industry. The notable exceptions are the Scandinavian countries.

In Sweden, collection is the responsibility of local utility companies run by the municipality. The first such company was set up in Gothenburg in

TABLE 8
National Data Management Systems (1)

	Austria	Denmark	FRG	France	Italy	Japan	Netherlands	Spain	Sweden	Southern Africa	UK	USA
Primary level of data collection	regional	regional/ national	regional/ national	regional	regional	regional/ national	national	no system at present	national/ local	no system at present	local	national
National data management system												
Is it computerized?	New	Yes	Yes	Yes	No	Yes	Yes		In progress		No	Yes
Is it regularly updated?	Soon	Partial	Yes	Yes	—	Yes	Yes		In progress		—	Yes
Date of most recent data	—	Yes	Yes	—	—	Yes	Yes		In progress		—	Partially
	(1983)	1985	1984	1986	None	1984	1984		1980		1979/85	1981/84
Basic methods of data collection					(2)							
Regular, mandatory report by producers	Yes	Yes	Partial	Yes		Yes	Yes		Yes		No	No

TABLE 8 (continued)

	Austria	Denmark	FRG	France	Italy	Japan	Netherlands	Spain	Sweden	Southern Africa	UK	USA
Report by treatment/ disposal (T&D) operators	Yes	Yes	Yes	Yes		No	Yes		Yes		No	Yes
Survey of trip tickets	Yes	Yes	Yes	Yes		—	Yes		No		No	No
Survey by national government												
- of industry/ T&D operators	Yes	Yes	No	No		No	—		Yes		No	Yes
- of regional/ local authorities	No	Yes	No	—		Yes	—		No		Yes	Yes
Survey of industry and T&D operators by regional/ local authorities	No	Yes	Yes	Yes		Partial	—		Yes		Partial	Yes

(1) See discussion in text.
(2) System still to be put in place.

2. SUMMARY AND ANALYSIS

1973 and by 1986 all municipalities had to arrange for the collection of hazardous waste (1). The utilities are responsible for collection and transport of hazardous waste, for the establishment of temporary storage facilities, and, in some cases, also for plants for pre-treatment of the waste.

In Denmark, there are some 25 central collection stations established through co-operation between the municipalities of an area. In addition, each of these stations serve some 300 smaller local collection points, at least one per municipality.

In general, responsibility for transporting waste to these central collection stations lies with the waste producer. However, some municipalities provide a collection service on request.

Recently, some municipalities have experimented with a less passive collection system, where they have gone to waste producers to collect hazardous waste without waiting for a request to do so. These experiments generally have shown significantly increased quantities of waste over the more traditional approach, an observation implying that signficant quantities of waste may be escaping control in countries where reliance is placed on the waste producer delivering waste for treatment or disposal. It is notable that, in both the Swedish and Danish country reports, emphasis is placed on the need to extend and improve their collection systems.

In most countries, hazardous waste is generally transported by road. The main exception is Denmark, where waste is transported from the regional collection centres to the central treatment and disposal plant primarily by rail.

Control of collection and transport

There are three basic methods that can be adopted for the control of collection and transport of hazardous waste:

- operating companies may be required to register with the authorities; or

- companies may be required to obtain a licence from the authorities; and/or

- a manifest or trip ticket system may be used to keep track of shipments of wastes, to ensure that they arrive at their designated destinations.

As previously discussed, registration is more common than licensing, but some countries do not adopt either.

There is more consensus as to the need for a manifest or trip-ticket system, a notable exception being Japan. However, the detailed working of

---

(1) Either by a municipal company, or by a private company engaged as a contractor on behalf of the municipality.

such a system varies considerably between countries, as summarized in Table 9.

Of the countries represented on the working group, eight have introduced a manifest system. In almost all of these cases:

- the manifest serves as an identification form that accompanies each shipment of waste;
- the manifest is signed at each stage of transfer of responsibility, from the producer to the collector, from the collector to the transporter (at a transfer station if one exists), and from the transporter to the treatment or disposal operator; and
- each operator in the chain of custody is required to keep a register, which is open to inspection by the authorities.

Beyond these basic elements, the details of the various national manifest systems differ significantly. In only half of the cases does the controlling authority get a copy of every manifest.

In only one case, the UK, does the controlling authority receive a copy of every manifest in advance of shipment, the purpose being to allow the authority to witness or supervise the treatment or disposal of the particular waste, should it wish to do so.

Not all countries use the manifest as a method of tracking individual shipments of waste to ensure that they arrive at their designated destination. Where this is done, the matching of manifests showing the dispatch and the receipt of specific shipments may be the responsibility of:

- the producer, who then notifies exceptions to the authorities;
- the authorities themselves; or
- where collection is the responsibility of a utility company, the central collection station.

Where "exception tracking" is practised, that is notification by the producer to the authorities if confirmation of receipt of the waste is not received, the delay involved in beginning the search for a "lost" waste is considerable, e.g. 45 days in the USA and 3 months in France.

The use of information from manifests for statistical purposes is currently unusual, although several countries are planning or investigating such systems.

In most countries, hazardous waste may move freely across internal boundaries within the country, for example between municipalities or between provinces or states. In the FRG, however, a special permit is required before a shipment of hazardous waste is allowed to cross state boundaries. Obtaining such permits takes a long time and is becoming increasingly difficult for political reasons.

TABLE 9
Characteristics of National Manifest Systems for Transport of Hazardous Waste (1)

	Austria	Denmark	FRG	France	Italy	Japan	Netherlands	Spain	Sweden	Southern Africa	UK	USA
General												
Is there a manifest system?	Yes	Yes	Yes	Yes	Yes	No	Yes	No	Planned	No	Yes	Yes
Date when introduced	1984	1974/75	1978	1985	1985	—	1980–	—	—	—	1981	1980
Record keeping												
Does identification form manifest accompany each shipment?	Yes	Yes	Yes	Yes	Yes	—	Yes	—	—	—	Yes	Yes
Is manifest signed at each stage?	Yes	Yes	Yes	Yes	Yes	—	No	—	—	—	Yes	Yes
Does each operator keep a register?	Yes	Yes	Yes	Yes	Yes	—	Yes	—	—	—	Yes	Yes
Tracking of wastes												
Does controlling authority get a copy of every manifest	Yes	Yes	Yes	(2)	No	—	Yes	—	—	—	Yes	No(3)

TABLE 9 (continued)

	Austria	Denmark	FRG	France	Italy	Japan	Netherlands	Spain	Sweden	Southern Africa	UK	USA
Does it receive copy in advance of shipment?	No	No	No	No	—	—	No	—	—	—	Yes	—
Are copies matched to track the waste?	Yes	Yes	Yes	Yes	No	—	Yes	—	—	—	Yes	Yes
Is this done:												
– by the producer who then notifies exceptions to the authorities?	—	—	Yes	Yes	—	—	—	—	—	—	—	Yes
– by the authorities?	Yes	—	—	—	—	—	Yes	—	—	—	Yes	No(3)
– by central collection station?	—	Yes	—	—	—	—	—	—	—	—	—	—
Information												
Are manifests used for statistical purposes?	Soon	Yes	Yes	Yes	No	No	Yes	—	—	—	No	No
Is system computerized?	Soon	Soon	Pilot scheme	Yes	—	—	Yes	—	—	—	—	—

(1) For explanations, see discussion in text.
(2) In France, the authority receives a periodical summary of manifests, primarily for statistical purposes.
(3) In the USA, some states receive copies of every manifest and use them to check each shipment.

2. SUMMARY AND ANALYSIS

Import and export of waste

International trade in hazardous waste has been the focus of attention of a number of international organizations in recent years, including the OECD, UNEP and the CEC.

Table 10 summarizes some information on transfrontier shipment of hazardous waste. The quantities involved are relatively small, although they are significant in certain European countries.

The main destinations for waste being shipped between countries include:

- specialized incineration or treatment facilities;
- incineration at sea; and
- landfills, either because local facilities are unavailable (e.g. in The Netherlands) or because another country offers commercially attractive terms (e.g. export from Western Europe to the GDR).

As shown in Table 2, most of the countries represented on the working group either have or will soon have control over both import and export of waste. This generally takes the form of a manifest system, including provision for prior notification of the authorities in the importing country.

Household hazardous waste

Legislation to control hazardous waste in many countries does not cover hazardous waste from households that is generally disposed of with municipal refuse. These wastes may cause problems, however, either in terms of contaminating leachate from landfill sites or of noxious emissions, in the form of dioxins, cadmium, or mercury, from incinerators.

One of the major priority areas highlighted in the individual country reports is that of removing such hazardous waste from household refuse. Two general approaches are being considered — the separate collection of hazardous components; and the re-design of consumer articles so as to eliminate hazardous chemicals at the source.

Many countries are beginning to institute, or experiment with, systems for the separate collection of household hazardous waste. In several countries, priority is being given to the development of general systems, aimed at the separate collection of e.g., batteries, paints, solvents, and waste pharmaceuticals. A number of different types of collection systems are being utilized.

In Sweden, some municipalities operate special scheduled collection services, while others collect packaged and labeled hazardous waste along with their regular weekly domestic refuse collection, so long as the collection is booked at least 1 week in advance.

In the FRG, some local authorities have instituted special collection

TABLE 10
Import-Export of Hazardous Waste

Country	Imports	Export
Austria	No national information. Included 1700 tonnes per year from FRG in 1980.	Mainly to landfills in the GDR and Hungary.
Denmark	Minor	1000 tonnes per year via the central treatment plant plus up to 11,600 tonnes per year exempted wastes.
FRG	40,000 tonnes per year in 1982, 34,000 tonnes per year in 1983. About 10,000 tonnes per year from The Netherlands for landfill, rest mainly for incineration, treatment or deep burial.	To the GDR for landfill: 1982 140,000 tonnes per year 1983 350,000 tonnes per year Also 40,000 tonnes per year mainly to France, Switzerland and Belgium.
France	100,000 tonnes per year for treatment, mainly from FRG, Switzerland and The Netherlands, plus 10,000 tonnes per year for landfill from The Netherlands	About 8000 tonnes per year
Italy	No national information. Included 4000 tonnes per year from FRG in 1983.	Estimated at 3500 tonnes per year
Japan	None	None.
Netherlands	Minor	About 120,000 tonnes per year in 1983, 50% for landfill. Main destinations for landfill are the GDR, Belgium, France and FRG; for treatment, UK, FRG and France. (Detailed statistics for 82/83 and 83/84 in country report).

TABLE 10 (continued)

Country	Imports	Export
Spain	Significant (but unknown) quantities for landfill	About 1000 tonnes per year for treatment.
Sweden	Permits needed from Jan 1, 1986. No information yet available. Less than 10,000 tonnes per year.	
Southern Africa	Some trade between countries, mainly exporting to cheaper landfills. Quantities unknown.	
UK	1984 5000 tonnes per year 1985 25,000 tonnes per year Low 1984 figure due to temporary embargo by one contractor. Almost all for incineration or treatment. Mainly from The Netherlands and Ireland, also Scandinavia.	Minor
USA	Minor.	Relatively small quantities to Canada and Mexico.

centers, which are open on a regular basis for the delivery of wastes by householders.

In the USA, six states and approximately 200 local communities have HHW programmes. The US EPA published a report in 1986, "A Summary of Household Hazardous Wastes and Related Collection Programs", which found participation in these programs to be low and costs per pound of hazardous waste collected to be high. It found these programs to be beneficial, however, in that the public became more aware of HHW and more likely to dispose of these materials in environmentally acceptable ways than prior to the collection programs.

Other country reports that highlighted these issues include those for The Netherlands, Denmark and Austria.

A particular focus of concern in a number of countries, notably Japan, is the separate collection of mercury batteries, and in some cases also other types of dry cell batteries containing mercury.

One approach to separate collection for consumer products, such as batteries or medicines, is to place special containers in retail shops.

Separate collection can be encouraged by the use of deposit schemes, where a deposit is repaid to the consumer on surrender of the used item. For example, in Austria, such a scheme has been considered in both mercury batteries and fluorescent tubes containing mercury. Denmark plans to introduce a deposit system for "household chemical" containers.

Another approach is the occasional "amnesty", where small generators or consumers are encouraged by advertising to return hazardous waste, or specific materials such as surplus medicines or pesticides, for a limited period. Examples of such schemes are in The Netherlands, the UK and some states of the USA.

The major problem with most schemes for the separate collection of hazardous waste is the low participation rate by the public. In the FRG, the efficiency of local authority schemes is estimated at 10–15%. In Japan, the recovery rate for mercury batteries is generally less than 20%, with a maximum of 40%.

In Denmark, participation rates are found to decrease after the initial period when publicity is high. Periodic publicity campaigns or the use of deposit systems are considered necessary to maintain participation.

In Austria, an interesting field test scheduled for 1 year has been carried out in two districts of Vienna. All household hazardous wastes could be deposited with mobile hazardous waste collection units. Reduction rates of approximately 50% were achieved on a continuing basis. As a result, a network of 33 collection units is currently being spread over the entire city area. The future development of this collection system will be watched with interest in other countries.

In the USA, the EPA report found that only 0.2 to 1.0% of total households participated in the established collection programs. The cost for these programs was between 2 and 9 US dollars per pound of waste collected.

In the long-term, the problem of household hazardous waste may partially be solved by the re-design of consumer products to eliminate the hazardous components. This has been highlighted as a particular priority area by three countries.

In Japan, manufacturers of dry-cell batteries have undertaken a program aimed at reducing the amount of mercury used in dry cells for the domestic market substantially over the next 3 years. In addition, the manufacturers have given 2 million US dollars (1985) to the Japan Waste Management Association to set up a research and development agency to investigate general problems in waste management.

In Denmark and Austria, consideration is being given to a scheme to label products containing hazardous materials, both to encourage householders to segregate them for separate collection and to give an incentive to manufacturers to use alternative technologies by enabling them to advertise it as a non-hazardous product. This is part of a wider initiative to label environmentally sound products.

Special collection systems

Most countries have regulations on waste oil, falling either under general legislation on waste, under legislation on hazardous waste, or in specific legislation.

In a number of countries, these regulations establish a collection system. In The Netherlands and Italy, private companies receive permits to collect waste oil, and are then obliged to provide a collection service if the quantity exceeds 200–500 liters. In Sweden, collection of waste oil is provided by the local utility company.

Most waste oil is used as a fuel, either with or without some preliminary cleaning or refining. Problems are being encountered in a number of countries, from contamination with chlorinated hydrocarbons, e.g. where polychlorinated biphenyl (PCB) containing transformer "oils" have been mixed with other waste oil. This leads to the possibility of severe air pollution if the waste oil is burned in small units where combustion may take place at a relatively low temperature. Bans on the burning of waste oil in small units, or of oil containing more than a specified amount of total chlorine, have been introduced or are under consideration in a number of countries, including Member States of the European Community as requested by an EC Directive, Sweden, and the USA. Such contaminated oil must then be handled as for other hazardous waste.

The Marpol Convention of 1973 governs marine pollution by ships on an international basis. Annex 1 of the Convention was introduced in October 1984, and requires that wash water from oil tankers should be discharged to tanks in port for subsequent treatment.

Many countries either already have or are in the process of introducing adequate facilities at their major ports (e.g. see the country reports on The Netherlands and France). If a ship complains that facilities at a particular port are inadequate, then the competent national authorities must investigate the complaint and ensure that facilities are brought up to standard. However, little attention has yet been paid to enforcement to ensure that tankers use the facilities provided.

Annex II of the Convention, covering wash waters from chemical tankers, come into force in April 1987.

The disposal of small quantities of a wide variety of chemicals from laboratories, particularly those in schools and other educational establishments, is a considerable problem. The most appropriate solution is probably collection by the local authority with subsequent sorting, bulking, and dispatch to the appropriate form of treatment or disposal. These problems are discussed in the country reports from France, the UK, and Sweden.

Technical problems in transportation

A number of technical aspects of transportation where current standards could be substantially improved were highlighted in the country report on France. Procedures for sampling and identifying wastes upon receipt at a treatment center are required, in order to verify that the waste conforms to its description.

Obtaining a representative sample of a generally heterogeneous waste is a difficult problem. This problem is particularly severe when wastes are transported in drums, the prevalent method in many countries. It is generally impractical to check the contents of every drum. It is for such reasons that the UK Hazardous Wastes Inspectorate, and authorities in other countries as well, have recommended that landfilling of drummed waste is unacceptable.

Washing trucks and tankers between loads is necessary to avoid mixing incompatible waste, both for safety reasons and because the mixture may be difficult to treat. The problem relates both to the provision of facilities and to ensuring that drivers take the time to use them.

2. SUMMARY AND ANALYSIS

STORAGE, TREATMENT AND DISPOSAL

Overview

The provision of adequate facilities for the storage, treatment and disposal of hazardous waste, measures to ensure their use, and the effective enforcement of proper controls, all form essential parts of a national control system.

The purpose of this section is to compare national systems for the storage, treatment and disposal of hazardous waste in more detail. It considers:

- the technologies used in the various countries, comparing their extent of use and highlighting significant differences both in policy and in practice;
- the extent of recycling;
- the approach to developing treatment and disposal facilities for the future, including such issues as the siting of new facilities; financial support for the development of both new waste treatment technologies and new recycling systems; and provision of research and development support to industry to solve their waste management problems; and
- problems of old and abandoned hazardous waste sites.

There was reasonable consensus among members of the working group as to the "hierarchy" of preferred management options for hazardous waste. Thus, for any particular hazardous waste stream, one should ideally consider the following options, listed in their order of preferance:

- waste avoidance or reduction at source;
- recycling, both energy and materials recovery;
- treatment or incineration to destroy, convert or immobilize hazardous constituents;
- disposal on land; and
- disposal on sea.

There was agreement that waste water following treatment may be acceptable for discharge to sewer or directly to surface waters. Despite this consensus, however, there was found to be considerable differences in emphasis in the policy of the different countries. This applies particularly to the legitimacy of land and sea disposal, that is whether or not these can be considered as acceptable options for any hazardous waste, and to how far economics should be taken into account in determining practicability.

Technologies for storage, treatment and disposal

The aim of this section is to examine the technologies used for storage, treatment and disposal of hazardous waste in the various countries, comparing their extent of use and highlighting significant differences both in policy and in practice.

The country reports contain much information on facilities available in different countries. However, in reading the following sections based on that information, a number of points should be borne in mind.

In a number of countries, notably France and The Netherlands, the availability of information on the treatment and disposal of wastes once they have left the place of generation is good. On the other hand, less tends to be known regarding treatment and disposal taking place within the producer's premises. For example, in The Netherlands, detailed statistics are available on treatment and disposal techniques, both within The Netherlands and broken down by importing countries, for rather less than 50% of the total hazardous waste, most of the remainder being handled in-house. The latter is a local, rather than a central, Government responsibility.

Differences in definition, as discussed previously, also make comparison difficult. Particular instances of this are highlighted in the text.

The following subsections examine a number of technologies in turn, beginning with storage, moving through incineration and other treatment methods, to landfill and other land disposal techniques, and finishing with disposal at sea. A final subsection attempts to make some limited quantitative comparisons of the extent of use of the different technologies in the various countries.

Storage

Some temporary storage of hazardous waste must occur at the producer's premises. In general, however, this is regarded as part of the production process and regulation is less stringently applied if at all. For example, specific control under hazardous waste legislation is only applied:

- if storage is for an extended period, in the UK defined as over 28 days and in the USA 90 days;
- if the storage is effectively more than a treatment process, such as a settlement pond for aqueous waste; or
- if significant mixing or blending of waste takes place, as in a tank farm.

In a number of countries, intermediate collection points or transfer stations exist, where waste is accumulated, stored, blended and sometimes pre-treated prior to shipment to a treatment or disposal facility. The most

2. SUMMARY AND ANALYSIS

extensive networks of this kind are in Denmark, Sweden and the state of Bavaria in the FRG. However, commercial waste brokers also exist in a number of countries, an example being The Netherlands where they particularly handle wastes destined for export.

In most countries, there will sometimes be the need for long-term storage of waste for which no obvious treatment or disposal option is available within the country. This is highlighted, for example, in the reports on Austria and Sweden.

Storage facilities are obviously required at incineration and treatment plants, to allow for the proper storage and blending of wastes, prior to their treatment. A useful technical discussion of storage facilities, at collection points, transfer stations and the central treatment plants, is contained in the Danish report.

Little information is available on numbers of storage facilities for hazardous waste in the different countries.

In the USA, some 4300 storage facilities were recorded in the 1981 survey: 85% utilized drums or vessels for storage, 33% used tanks and 12% used surface impoundments. In terms of volumes stored, the latter two methods predominate.

In the UK, some 100 plants are licensed for storage. This excludes both temporary storage, for less than 28 days, and those plants whose license includes treatment or incineration as well as storage.

This section has considered temporary storage. Permanent storage of hazardous wastes, for example in salt mines, is regarded as disposal, and is considered in detail later in the report.

Incineration

Incineration is the high temperature combustion of hazardous waste in the presence of a controlled excess of air. If the combustion gases are held at high temperature for sufficient time, in principle all of the organic materials should be completely oxidized, mainly to carbon dioxide and water vapor, while inorganic components should remain in the ash.

Incineration plants are often an integral part of the production process, particularly in the chemical industry. A number of industrial plants also have their own, more general purpose incinerators for treating wastes from several processes.

All of the countries represented on the working group that have instituted comprehensive controls on hazardous waste also have one or more centralized incineration facilities available on commercial basis to any waste producer. The numbers of commercial facilities vary, examples being one each in Austria, Denmark, The Netherlands and Sweden, four in the UK, 10 in France, 17 in the FRG and 120 in Japan.

Numbers of in-house incinerators are more difficult to gauge: there are about 60 in the UK and 360 in Japan, while the total number of in-house and commercial incinerators in the US is about 230.

The most popular type of incinerator for general purpose, centralized facilities accepting a range of solids, sludges and liquids is the rotary kiln. A slowly rotating drum moves solids through the process, sufficient gas retention time normally being provided by a separate afterburner chamber. Rotary kilns are also used at a number of in-house facilities in the FRG, while a total number of 42 such facilities were reported in the USA in 1981.

The most common type of incinerator, particularly for in-house facilities, is probably the liquid injection incinerator, which, as its name implies, is limited largely to pumpable liquids as feedstock. Three of the four commercial incinerators in the UK are based on liquid injection designs, but two have been modified to accept either sludges or solids in addition. In the USA in 1981 there were some 219 liquid injection incinerators.

A third generic type of incinerator for hazardous wastes is the fluidized bed, which again is limited to liquids or pumpable sludges. The extent of use of such technology is fairly limited, although a few such incinerators are known to exist in the USA.

An interesting development is the use of cement kilns for hazardous waste incineration. This has the threefold attraction that:

- heavy capital investment is not required;
- the alkaline cement materials neutralize acid gases from incineration; and
- the ash from incineration is incorporated within the cement product, so it does not require disposal.

The extent of use of cement kilns for hazardous waste is still limited. A few facilities in the USA are currently attracting almost all kinds of hazardous wastes. In France, five kilns burn about 20,000 tonnes per annum, largely comprising acid tars otherwise difficult to dispose of. In Italy, waste oil is utilized as a fuel in cement kilns.

Some hazardous wastes also are burnt as a fuel in industrial boilers. This applies particularly to waste oils, some of which may be contaminated with other materials. Strict controls on such incineration are being introduced in a number of countries.

Co-incineration in municipal waste incinerators is used in a number of countries, particularly for small quantities, for example of packaged laboratory waste. About 50% of the hazardous waste incinerated in The Netherlands, that component requiring a relatively low temperature for combustion, is handled by municipal incinerators. There are 25 municipal or industrial incinerators licensed for co-combustion of small quantities in the UK.

Calcination is a variation of the incineration process, where the aim is to convert organic metal compounds into the inorganic form. This is used particularly for leaded gasoline sludges or, notably in Japan, for sludges containing organo-mercury compounds. The design of such incinerators is fundamentally different, as the temperature must be high enough to convert the organic metal compounds to inorganic forms, but must not be so high that the metal is lost as fume or vapor with the off-gases.

Research is under way in a number of countries, notably the USA and the FRG, to develop new, advanced incineration methods. These include high temperature incineration and the use of electrically heated furnaces.

Incineration of hazardous wastes has the potential for environmental pollution, either via the off-gases produced or via the inerts remaining for disposal. Removal of particulate materials from the gases is essential, while, if the waste contains more than a minimum percentage of certain elements, principally chlorine, nitrogen and sulfur, the acid gases also must be removed, generally by a wet scrubber, which produces an aqueous effluent that needs to be treated prior to disposal.

Air pollution control is cited in a number of country reports as a critical problem facing incineration. Particular concern focuses on the possible production of trace chemicals such as dioxins, which are both highly toxic and highly emotive. Much scientific work has addressed this question, with the general conclusion that, given proper design and control of the incinerator, the risk is minimal.

Of the total number of hazardous waste incinerators, relatively few are equipped with the air pollution control equipment necessary to burn highly chlorinated waste. A particular problem relates to PCBs, in that some incinerators technically capable of burning these materials are not licensed to do so. Most PCB waste arising in western Europe is currently incinerated at four plants in West Germany, two in France, two in the UK, and one in Sweden. The first three countries all import substantial quantities for incineration. The new plant in The Netherlands also will accept PCB wastes beginning in 1987.

The ash remaining from incineration of hazardous waste is generally disposed of in landfills. As discussed previously, there are differences between countries as to whether this is regarded as a hazardous waste or not.

Incineration of high calorific value wastes may be regarded as a form of recycling if the heat generated is utilized. Most plants use heat internally, but external sale of energy is practised at a number of plants, including the central treatment plants in Austria, Denmark, Sweden, and the FRG, and at three out of the ten plants in France.

There are currently three specialist incinerator ships installed with liquid injection incinerators operating at a temperature around 1500°C. This is significantly higher than the 1000–1200°C common in land based incinerators. This system is used primarily for chlorinated wastes, the plume of acid gas being allowed to settle out into the sea.

A number of extensive scientific studies have been made on incineration at sea, from which the general conclusion has been that, from a technical point of view, it is successful and has minimum effect on the environment.

Incineration at sea is controlled under both the London Dumping Convention (worldwide) and the Oslo Convention (the North Sea and North-East Atlantic). The signatories to the Oslo Convention have agreed to phase out incineration at sea, subject to the results of a new series of scientific trials to be undertaken in 1987 (these tests have now been cancelled, so the position is unclear).

The total quantity of wastes incinerated at sea from Europe was around 100,000 ($\pm$ 10,000) tonnes per annum between 1979–84. The composition of this waste is such that current land-based incineration capacity is insufficient to deal with it.

Physical and chemical treatment
The pattern of physical and chemical treatment is similar to that for incineration, with facilities being provided both in-house and at centralized, commercial facilities in those countries with more developed control systems.

The most common treatment processes are:

- thickening and de-watering of sludges;
- neutralization of acids and alkalis;
- precipitation, for example of heavy metals;
- reduction of chromium (VI) to chromium (III);
- oxidation of cyanide; and
- solidification of sludges.

The majority of treatment facilities operate a single process or a limited range of processes. A small number of the commercially available sites operate a range of processes that may also include oil-water separation, discussed here as a recovery process.

The general principle of most treatment methods is to produce an aqueous waste acceptable for discharge to sewer or to surface waters, and a dewatered sludge or solid waste acceptable for landfilling. Technical problems are usually associated with achieving the appropriate quality standards for the residual waste.

Detailed information on the numbers of treatments plants of different types is given in the country report on Japan. The following information applies to industrial (rather than hazardous) waste treatment plants in 1985.

Type of Process	Approximate number
Sludge dewatering	4800
Sludge drying	170
Acid or alkali neutralization	240
Cyanide oxidation	310
Concrete solidification	100

These figures illustrate the problems in using comparative statistics. Many factories have sludge dewatering plants as part of the process, but they will not necessarily be included in national industrial or hazardous waste statistics.

Another interesting point from the Japanese figures is the relatively large number of plants for concrete solidification. About 40% of these facilities are in-house, the remainder being commercially available. (This is in contrast to cyanide oxidation plants, of which more than 80% are in-house.) In most other countries, solidification of inorganic sludges takes place at a few commercial plants, using proprietary processes, usually based on a combination of cement and an inorganic waste such as fly-ash.

Research is proceeding in a number of countries on new treatment processes for hazardous wastes. This is particularly the case in the USA where a number of wastes will soon be banned from land disposal. Examples of new processes include:

- dechlorination to detoxify persistent chlorinated substances; and

- an extension of chemical oxidation to organic cyanides, phenols and organic sulfur compounds.

In the UK report, a new process is reported for the conversion of fibrous asbestos to a harmless glass in a specially adapted high temperature furnace. Research in the FRG focuses on chemical and thermal destruction of highly chlorinated hydrocarbons.

Landfill

In a number of countries, both public perception of, and Government policy towards, landfill disposal of hazardous waste is colored by the widespread occurrence of old and abandoned problem sites. Such problems are attributed primarily to the uncontrolled mixing of hazardous with other waste and to the leakage of chemicals into groundwater.

As a result of this past experience, policy in a number of countries favors an engineering concept of landfill, where hazardous waste is segregated in a site selected or engineered so as to isolate the waste as far as possible from the environment. Any leachate generated by the waste is collected and treated, and both ground and surface water in the vicinity of the site are monitored.

In some countries, notably the USA, technical standards for site construction, e.g. in terms of providing a double liner, an impermeable cover, leachate treatment, and groundwater treatment, required for at least 30 years after closure. Insurance cover to pay for any necessary corrective action is a current issue in a number of countries, and is already mandatory in the USA.

There is an increasing trend to restrict the input of hazardous waste to such landfill sites largely to residues from incineration or chemical treatment plants.

This oversimplified description broadly characterizes the situation in Denmark, the FRG, Sweden and the USA. Legislation in Austria and in Italy points in a similar direction, but engineered landfill sites have yet to be provided. In France, there are 13 such controlled landfills, but there are a further 80 in-house landfills on which little information is available. In addition, 20% of all hazardous waste in France is estimated to go to other landfill sites, with less than the ideal controls.

The UK is well-known for its advocacy of an alternative concept of landfill, controlled co-disposal of hazardous waste in municipal waste landfill sites. It is argued that properly controlled co-disposal of selected hazardous waste up to a maximum loading rate can result in the degradation of certain organic contaminants or the attenuation of inorganic contaminants to background levels. In either case, the ideal result is that the waste may no longer be hazardous. The proponents of co-disposal thus claim that it is less likely to lead to future problems of contaminated sites than the alternative policy of deliberately segregating and entombing hazardous waste, which in effect is a policy of controlled long-term storage. The opponents of co-disposal are sceptical that reliance can be placed on processes within a landfill for primary control of hazardous waste, viewing it rather as a means of spreading contamination over more sites.

Considerable research has been conducted in the UK on the scientific basis for co-disposal, and comprehensive guidance recently has been published both on control procedures and acceptable loading rates. However, as made abundantly clear in the first two annual reports by the new Hazardous Waste Inspectorate, the controls over co-disposal required in the policy are not yet being enforced in practice.

In a number of countries represented on the working group, notably Italy, Spain and Southern Africa, where controls over hazardous waste are either recent or not yet introduced, general practice is for the uncontrolled disposal of hazardous waste on municipal or other landfill sites. Some municipalities in Southern Africa operate impermeable landfill sites for the controlled co-disposal of hazardous with municipal waste.

Two countries have so far not been mentioned in this review. In The Netherlands, landfill of hazardous waste is prohibited unless specific exemption is granted. About 12,000 tonnes per annum are landfilled under such exemptions, mainly in co-disposal sites. A new engineered landfill site for chemical wastes is under construction and should be in operation by 1988. Currently, about 60,000 tonnes per annum of hazardous wastes are exported for landfilling in other countries, more than 50% to the GDR.

The Netherlands is in a unique position, having an almost total absence of suitable sites within its territory to landfill either hazardous or municipal waste. It is to be expected thus, that large quantities will be exported for landfill, at least until 1988 when the concrete-lined landfill is in operation. However, export for landfill also occurs from other countries, notably Austria and the FRG. The quantities exported from the FRG and the GDR have increased rapidly since 1982.

In Japan, three catogries of landfill sites are distinguished. Type I facilities are engineered, concrete-lined pits, which are covered during their lifetime by plastic sheeting to prevent rainfall infiltration, and are finished off with concrete on top. Type II sites are engineered sites with leachate collection systems, similar to the concept of secure landfill in other countries. Type III sites have no leachate protection measures.

By law, hazardous waste, under the Japanese definition, can only be disposed of in Type I landfill sites. There are 30 of these sites in the country, compared to 800 Type II and 500 Type III sites.

While the total number of landfill sites in some countries may be relatively large, the number of sites used for hazardous waste disposal is relatively small. In addition to the one site in The Netherlands, 13 in France and 30 in Japan quoted above, there is one in Denmark and 22 in the FRG. In the UK, a relatively large number of sites are licensed to take some hazardous waste, but only 35 receive more than 5000 tonnes per annum. Of the larger number of licensed sites, about 50% are only licensed to receive asbestos, for which controlled co-disposal in municipal waste landfill sites is standard practice in all countries (1).

Surface impoundments

Aqueous waste may be treated in surface impoundments such as pits, ponds and lagoons. This may be regarded either as storage or as a form of waste water treatment, allowing the settling of solids and perhaps some biological degradation prior to discharge of effluents to surface waters.

Surface impoundments are used primarily in the USA, where they are now subject to regulation along with other forms of land disposal. In 1981, 19 million tonnes of hazardous waste were disposed of in surface impoundments. This is in addition to the 62 million metric tonnes treated in

---

(1) Both blue and other fibrous or dusty asbestos wastes are defined as "special" wastes in the UK, this being a more restrictive definition than that for hazardous waste. However, asbestos is not controlled as a hazardous waste in certain other countries, notably the USA.

surface impoundments and the 52 million metric tonnes stored in surface impoundments (1).

Surface impoundments are also used in Spain, Southern Africa and the UK. Elsewhere, use appears to be limited, to, for example, canal dredgings.

After treatment in surface impoundments, waste water treatment or physical-chemical treatment facilities, the aqueous effluents are generally discharged to sewer or direct to surface waters. In most countries such effluents are not regarded as hazardous wastes, an exception being the USA.

Land farms

Land-farming comprises the mixing of sludges into the top surface of the soil, usually by ploughing. The method was developed and is successfully used for refinery sludges in the USA, but it also has found some application in Europe. In 1981, there are some 70 licensed land treatment sites in the USA, treating 400,000 tonnes per annum of waste. Since 1984 these sites have been brought under similar controls to other land disposal sites. In the UK, there are nine licensed land farm sites.

Land-farming of oil sludges is a treatment method, aimed at the biological breakdown of the sludge. The land is not used for agriculture. This is in contrast to the application of sewage and other biological treatment sludges to farm land, a common practice in a number of countries. Control is necessary when the sludges are thought to be contaminated with heavy metals. This is an area that has been much studied, and is beyond the scope of the working group reports.

Subsurface disposal

Three forms of subsurface disposal may be distinguished:

- deep-well injection;
- disposal of liquid wastes via mine shafts; and
- deep burial.

Deep-well injection of hazardous waste is confined, among members of the working group, to the USA. Some 30 million tonnes of aqueous waste, 11% of all hazardous waste, were disposed of by this method in 1981. However, under new regulations this disposal method is expected to be phased out for certain wastes.

Disposal of liquid waste to mine shafts is practised to a limited extent in the UK. There are a total of 14 licensed facilities, mostly in-house facilities utilized for particular types of wastes. They include the re-injection of saturated brine and certain other liquid wastes to salt cavities. The major

---

(1) These amounts, incidentally, total 133 million tonnes approximately half of the 264 million tonnes produced in 1981.

2. SUMMARY AND ANALYSIS

commercially available facility disposes of aqueous waste contaminated with organic materials in a mine shaft purported to be totally sealed.

In Spain, a gypsum mine is used for the disposal of dewatered residues from chemical treatment.

Deep burial of waste is exemplified by the long-term storage of containerized wastes in salt mines. The major commercial facility, in the FRG, receives about 40,000 tonnes per annum of particularly difficult and toxic wastes, from most European countries. Each individual shipment has to be permitted by the state government of Hessen. The location of each consignment within the mine is recorded, so that waste can be excavated later for recycling if appropriate.

Ocean disposal

Dumping of waste at sea is governed by two international conventions, the London Dumping Convention applying worldwide and the Oslo Convention to the North Sea and the North East Atlantic.

Under these conventions, it is an offence to dump in the sea, or to load for the purpose of dumping, any material without a license from the competent authority within the country of origin, and except in accordance with the conditions of the license. Both conventions specify in detail the factors to be considered before a license is granted, and in addition list materials for which dumping should be prohibited and others for which special care is required.

The prohibited substances are mercury, cadmium, and their compounds, organohalogen compounds, some carcinogens, persistent floatable plastics and (Oslo only) organosilicon compounds. The materials to be treated with special care include wastes containing arsenic, lead, copper, zinc, cyanides, fluorides and pesticides.

Shallow sea dumping is utilized in a number of countries. National statistics are available, but the country reports are generally weak in this area. An exception is the UK, where 260,000 tonnes were disposed to sea in 1985. Data for 1975-81 are available in an International Maritime Organisation (London Convention Secretariat) report (1).

As with incineration at sea, some countries are working towards a complete ban on ocean dumping of industrial wastes. Denmark stopped dumping in 1982. The Federal Republic of Germany has announced that all dumping will stop by 1989, while a draft directive under discussion by the European Community would enforce a similar ban on all 12 member states.

Extent of use of the technologies

Information from the individual country reports on the approximate numbers of facilities for hazardous waste treatment and disposal by the various methods outlined in this section is summarized in Table 11. There

---

(1) Report on the nature and quantities of wastes dumped at sea. International Maritime Organisation Report no. LDG/SG.8/INF.4, 20 December 1984.

TABLE 11
Approximate Numbers of Facilities for Hazardous Waste Treatment and Disposal (1)

Technology	Austria	Denmark	FRG	France	Italy	Japan(2)	Netherlands	Spain	Sweden	Southern Africa	UK	USA
Incineration	1/−	1/*	17	10/+	3/+	160/360(3)	1/+	−/*	1/+	−/+	4/60	230
Physical-chemical treatment	+/+	1/*	23	8/+	*	1200/4500(4)	5	2/+	2/+	2/+	16/24	1000(9)
Landfill	(+)/(+)	1/−	22	13/80(5)	*	10/20(6)	−(7)	+	4/+	3/+	35(8)	200
Surface impoundments	−	−	−	−	−	−	−	+	−	2/+	50	1080(10)
Mineshaft disposal	−	−	−	−	−	−	−	1	−	−	14	−
Deep burial	−	−	1	−	−	−	−	−	−	−	−	−
Deep well injection	−	−	−	−	−	−	*	−	−	−	−	90
Landfarming	−	−	−	−	−	−	2	−	−	(+)	9	70

x/y x commercial (publicly available) facilities plus y in-house facilities
+   facilities known to exist, number not known
(+) facilities exist, but of little importance
−   no facilities
*   no information available

(1) All numbers rounded to two significant figures.
(2) Japanese data is for industrial wastes, 1985.
(3) Sludge incineration plants only. Excludes incineration of waste oil (150/300) and of plastic (400/950).
(4) Breakdown of physical and chemical treatment methods in Japan is given in text.
(5) Excludes 20% of hazardous waste deposited at unauthorized sites.
(6) Type I landfill sites (concrete lined pits).
(7) One landfill site under construction. Landfill of specific wastes permitted as an exception at a number of sites.
(8) Sites receiving more than 5000 tonnes per year of hazardous wastes.
(9) Includes only treatment in tanks and tank-like devices. Treatment in surface impoundments is not included.
(10) Includes 120 disposal impoundments, 550 storage impoundments, and 410 treatment impoundments.

2. SUMMARY AND ANALYSIS

are a number of places in the table where information is not available or where a method is known to be used but no specific information on numbers of facilities is available. However, the overall availability of information is quite good, and the various trends discussed already in the text are apparent.

Unfortunately, when one tries to go beyond the number of facilities to either quantities or percentages treated by each of the methods, then a number of the problems discussed earlier reappear. For example:

- there are differences in definitions of hazardous waste between countries;
- each country that produces statistics chooses a different grouping of options for presentation of the material;
- statistics for France and the FRG exclude disposal at sea, the inclusion of which makes a considerable difference to percentages; and
- a breakdown is only available in The Netherlands for those wastes that leave the producers premises, amounting to rather less than half of the total.

Waste recycling and avoidance

Oils, solvents and waste heat recovery

The recovery of waste oils, solvents and waste heat from incinerators is commonplace in a number of countries.

Oils or solvents can generally be treated either by refining or distillation, to produce a marketable product; simple processing to produce a fuel product; or direct combustion for its energy value.

In addition to processing of waste oil, many plants exist for the separation of oil-water mixtures. Numerous techniques are available depending on the characteristic of the waste. A recent development is the evapo-incineration process for soluble oils developed in France.

Statistics are available in a limited number of countries for the quantities of waste oil and solvents recycled by the commercial sector. However, particularly for solvents, much larger quantities may be recycled in-house. An example of this is provided by the following statistics for France, Table 12, on the usage of solvents.

TABLE 12

	Million tonnes per year
Solvent usage	2.8
Used in gas phase and recycled	1.5
New solvents	0.8
Regenerated in-house	0.4
Recycled by the commercial sector	0.05

A similar situation is thought to exist in the UK, where 200,000 tonnes per annum are recycled by the commercial sector. In the USA, 80% of solvent recycling is thought to take place in-house.

Recycling is not without its own environmental problems. Distillation of solvents leaves still-bottom residues, which often require incineration using a support fuel for their proper disposal. Thus, at a certain contamination level, direct incineration of the original solvent waste becomes more cost-effective. In the UK, the Hazardous Waste Inspectorate has expressed concern at the activities of small solvent recoverers who are handling these "uneconomic" wastes, presumably using improper disposal methods for the residues.

Heavy metals
Commercial and in-house operations exist in a number of countries to recycle heavy metals from various sources:

- Silver from photo-finishing operations.

- Lead from lead-acid batteries. Like many recovery processes, this requires strict control, as the residues from the processing are themselves hazardous wastes.

- Heavy metals such as copper, nickel and zinc from metal finishing wastes, particularly plating baths, from catalysts and from the iron and steel industries. An innovative process in Sweden uses a plasma reactor to concentrate metals from metal finishing sludges.

- Precious metals, for example from computer debris.

- Mercury, from batteries, dental amalgam, broken thermometers and the chlor-alkali industry. Mercury recycling plants are known to operate in France, the FRG, Italy, Sweden and the UK. There is an overall shortage of capacity and considerable international trade occurs. The rate of recycling is quoted as 27% in the FRG. Further discussion of the subject is given in the report on France.

Despite the widespread occurrence of recycling, the overall quantities of heavy metals recycled is relatively low.

Waste avoidance and recycling within industry
Waste recycling is an integral part of many industrial processes. A basic economic principle of industry is to maximize the proportion of raw material converted to useful products. Thus, by-products from one process are commonly used as the raw material for another. What is today an innovative recycling process will soon become an accepted part of the technology, so that it is generally not possible to obtain any meaningful information on the rate of recycling in a particular industry.

2. SUMMARY AND ANALYSIS

A number of examples serve to illustrate the technical possibilities for increased recycling. Sulfuric acid may be regenerated from the dilute acid resulting from titanium dioxide production. At least one such plant has been in operation in the FRG for a number of years and a second will start in 1989, when discharging at sea will finish (deadline set by the Federal Government).

A number of processes exist for the recovery of chlorine values from highly chlorinated wastes resulting from the production of, for example, PVC and dry-cleaning solvents. Such wastes now form the prime feedstock for the manufacture of carbon tetrachloride and tetrachloroethylene in a number of countries. An alternative is to recover hydrogen chloride from incineration. Of 600,000 tonnes of such wastes produced in the European Community, about 400,000 tonnes are recycled.

Manufacture of phosphoric acid (as an intermediate in fertilizer production) produces large quantities of phosphogypsum wastes. In most countries, this is dumped, either on land or at sea. In the European Community, about 10% is utilized in building materials. In Austria, phosphogypsum is used as a raw material for making both ammonium sulfate and sulfuric acid. A new process developed in Denmark results in the production of soluble potassium sulfate rather than insoluble gypsum.

Two examples for the reduction of hazardous waste by changing products are provided by the substitution of solvent-based by water-based paints (Sweden, Denmark) and the replacement of mercury in batteries (Japan, Sweden). Additional examples of recycling processes or of low waste technologies are given in the Austrian and Japanese reports. In many countries, it is official policy to encourage new technologies for waste avoidance or recycling.

Waste exchanges

A waste exchange scheme exists to promote the use of one company's by-product as another's raw material. Such schemes may be either passive, simply publishing lists of waste available and waste wanted and allowing interested parties to make contact via the exchange, or active, where professional staff employed by the exchange actively seek outlets for waste or sources for materials required.

In a number of countries, waste exchange schemes have not "taken off". It appears to be difficult for such schemes to be commercially self-supporting. Exchanges still exist in a number of countries, but these focus primarily on commercial and industrial rather than hazardous waste.

There are some examples of small local waste exchanges for hazardous waste that have proven successful. These tend to be active schemes, either provided by a controlling authority (example in the UK report) or by a commercial company (examples in Italy and USA).

Future recycling, treatment and disposal facilities

So far the focus has been on the existing situation. In this section, attention is turned towards developing facilities in the future.

Many countries report a lack of adequate landfill sites and a shortage of treatment capacity for hazardous waste. Progress in overcoming the immediate shortages is often slow due to problems in siting new facilities. Different national approaches to the siting problem are discussed below.

In the longer term, most countries wish to promote development of new technologies, for avoiding waste production, for recycling and for treatment and disposal. In this section, alternative approaches to providing both financial and technical support for such developments are reviewed.

Both the provision of adequate treatment and disposal facilities, and the encouragement of new technologies for the longer term, form part of a coherent national strategy or plan for hazardous waste management. As noted earlier, the need for such plans is recognized in a number of countries, but progress in actually producing the plans has often been poor.

Siting new facilities

Siting new hazardous waste facilities is made both slow and difficult in a number of countries by the almost inevitable opposition of local residents — the well known NIMBY or "not in my back yard" syndrome.

An interesting international comparison is that of arbitation procedures to decide between local opposition to a facility and the "national interest". In Italy, the local municipality has an effective right of veto.

In the USA, siting decisions are often controlled by the state governments, who can sometimes override local opposition. However, in the case of a nationally important facility, the federal government has no power to override the veto of a particular state.

In a number of countries, any dispute between the proponents of a national facility and the local municipality is referred to a tribunal. This is essentially the case in Sweden, Denmark, and the FRG, The Netherlands and the UK and some states in the USA, although the details of the procedures vary widely. For example:

- in some cases the decision of the tribunal is final;
- in Sweden, a local veto still applies in some cases but not in others; while
- in the France and the UK, the role of the tribunal is advisory, the final decision resting with regional (France) or central (UK) government.

Financial support for low waste and recycling technologies

In many countries, the primary means for encouraging waste avoidance or recycling is the imposition of strict controls on hazardous waste disposal,

2. SUMMARY AND ANALYSIS

accompanied by the charging of a realistic price (that is the implementation of the Polluter Pays Principle).

However, a number of countries have supplemented such controls and fee systems with limited subsidies to encourage industrial investment in pollution control. The type of financial aid available has included grants, soft loans, including loans at concessionary interest rates, and special tax allowances.

The subsidies have been available for capital investment, research, development, demonstration projects, and disseminating information.

The majority of pollution subsidy schemes have been directed toward water and air pollution, with only a minority aimed at the support of cleaner technologies and the recycling or re-use of residues. Most schemes apply only to new technologies, rather than to transfer to existing technology.

In Austria, the Environmental Protection Fund provides grants, low interest loans or financing of borrowing costs for "collection, recycling and disposal of hazardous waste". The aim is to encourage the replacement of existing polluting technologies. In 1985, the fund had about 70 million US dollars at its disposal.

In the FRG, two schemes have been operational since 1976 to give financial support to any project related to the avoidance of pollution. Support for research and development is available from the Ministry of Technology and Research, with funds specifically set aside for the area of solid waste. Capital grants also are available from the Ministry of the Interior to support plant modifications aimed at reducing pollution.

Under the November 1986 amendment to the FRG Waste Disposal Law, the emphasis for these programs will move from the avoidance of pollution to the avoidance and re-utilization of waste.

In Denmark, the National Agency for Environmental Protection introduced grants for recycling and cleaner technologies in 1984. While recycling grants cover both research and development, and capital investments, grants for cleaner technologies at present (September 1986) are only given for research, development and information projects. In both cases, grants are available only for innovative technology.

In The Netherlands, financial support for recycling and cleaner technologies has been available for more than 10 years. Funds may be given for initial market research, for pilot plants, for development work and for demonstration projects.

In the USA, financial support (from the Superfund tax) is available for the demonstration of innovative technologies for treating or destroying waste from abandoned sites. A product of this program is a mobile incinerator that can be transported to an actual site for use in site renovation.

In a number of other countries, support for research and development in recycling is also available through central government. However, such monies usually come from general research funds, with no specific budget set aside for the support of recycling.

Technical support for waste avoidance and recycling
Development of new procedures for avoiding the production of waste or for recycling may be constrained, within individual companies, by the limited availability of management time and of research and development resources, for which there are many other, competing demands.

It may thus be argued that it is effective for governments to provide technical support to industry as well as direct financial support. For example, a government agency to promote waste reduction and recycling could:

- undertake or sponsor research and development;
- provide an information service, and disseminate research results; and
- provide advice to companies with specific problems.

Every scheme to provide such technical support differs in its details. In France there is a national agency, ANRED, for support of research and development in some aspects of waste management. Individual treatment centers also carry out research.

In Austria, the Environmental Protection Fund provides expert advice on the selection of environmentally sound technologies in addition to its financing function. In Denmark, the national utility company acts as a consultant, helping the generator to overcome problems.

In one major metropolitan area of the UK, the local controlling authority operates an "active" waste exchange. Information generated by their statutory waste disposal duties is used to assist local users for particular, environmentally "significant" wastes.

In Italy, a commercial "waste brokerage" company will advise industry on process modifications to make wastes more recyclable, and will search out the best users for each by-product.

In the USA, there is at least one "hazardous waste extension service", operating out of a university. The aim is to help business employing up to 50 people to develop individual plans for the handling of hazardous waste. The service is confidential, with no threat of involving the regulatory agencies.

In Japan, the battery manufactures have given a grant to the Japan Waste Management Association to set up a research and development agency to investigate problems in hazardous waste management.

2. SUMMARY AND ANALYSIS

Problems of old or abandoned hazardous waste sites

In a number of countries, the term "hazardous waste" is synonymous in the public mind with the problems of old or abandoned sites and other sites where proper controls have not been implemented. This is particularly the case in the USA and in certain European countries including Denmark, The Netherlands and, to a lesser extent, the FRG.

In all of these countries, there have been well publicised examples of abandoned hazardous waste sites that have caused serious groundwater pollution or public health problems. Perhaps the most famous of these are at Love Canal in the USA and at Lekkerkerk in The Netherlands.

In several Euorpean countries, notably the UK, France and the FRG, abandoned hazardous waste sites are seen as just one part of a wider problem of industrially contaminated land. The closure of many traditional industries, often located in inner city areas, has released many thousands of hectares of land in need of redevelopment for a beneficial use. Often this land has been contaminated by the polluting nature of industries, including the storage or disposal of waste materials within the area of the factory. Typical examples of such industries include the manufacture of town-gas, iron and steel, coal storage and docklands.

The remainder of this section focuses on three particular aspects of the problem of abandoned sites:

- the compilation of national inventories of such sites;
- remedial programmes for clean-up; and
- provision of finance.

National inventories

In a number of countries, a systematic search has been made for abandoned waste disposal sites containing hazardous wastes. Available statistics are compared in Table 13.

In addition to these countries, national inventories have been carried out in Austria, the FRG and France (see Table 2).

TABLE 13
National Statistics on Uncontrolled Hazardous Waste Sites

Country	Total number of waste sites	Total number containing hazardous waste	Sites requiring immediate action
Denmark	3100	900	110
Netherlands	—	4350	1000
Sweden	3900	500	21
USA	—	20,000	2000

Remedial programs

Extensive, systematic remedial programs, focusing on priority sites identified in the national inventory, are in progress in Denmark, The Netherlands and the USA, while a similar scheme is beginning in Sweden. Although there are no special programs in either the FRG or France, significant numbers of sites have been cleaned up within the last few years.

The type of clean-up varies from site to site and also from country to country, the latter being partly due to differences in circumstances and partly due to differences in policy.

The remedial program at national priority sites in the USA is strictly regulated under national legislation. The country report contains an extensive discussion.

There are considerable technical problems involved in cleaning soil from abandoned hazardous waste sites. These include a general shortage of capacity for the treatment and disposal of contaminated soils, particularly those contaminated with organic materials including dioxins. The clean-up of contaminated groundwater is another area receiving much attention.

Financing clean-up programs

In a number of countries, a special fund has been set up to pay for remedial actions at abandoned hazardous waste sites. In some cases, for example in The Netherlands, the fund comes from general taxation. In others, for example in the USA, the majority of the fund comes from a tax on chemicals and petroleum. In both of these example countries, powers exist to force a "responsible party" to pay for clean-up, if that party still exists. In the USA, a producer whose wastes went to a particular site remains responsible, even if other waste disposers evade paying their shares for clean-up through bankruptcy or other forms of legal action resulting in the disappearance of firms and/or individuals responsible.

Provisions for financing the clean-up of abandoned sites in the various countries are summarized in Table 14.

A related topic is financial security of present facilities to ensure that adequate funds are available to clean-up any pollution that might occur in the future. Such arrangements are already required in the USA. In other countries, mechanisms for providing such financial security are being investigated. A particular problem is that Environmental Impairment Liability (EIL) insurance is seen as a long-term, high-risk business of little interest to the insurance industry at a time of excess demand for their services.

TABLE 14
National Sources of Finance for the Clean-up of Abandoned Hazardous Waste Sites

Country	Comments
Austria	No specific provision.
Denmark	Government funding. Costs estimated at least 50 million dollars over 10 years.
France	A special fund is being initiated voluntarily by the chemical, oil, steel and automobile industries and the waste disposal authorities, as an alternative to imposing a special tax.
FRG	No specific provisions have yet been made. Discussions are under way between government and Confederation of German Industry.
Italy	No government assistance is available.
Japan	Public funds have been used in specific cases.
Netherlands	A fund of about 1700 million dollars has been provided to clean-up abandoned sites. National Government pays 90% and local government 10% of clean-up costs. If the polluter still exists he is charged for the work.
Sweden	No specific funding has been made available yet but it is being discussed.
Southern Africa	Such clean-up as has been carried out has been funded from general taxation.
UK	No specific provision.
USA	The original "Superfund" allocated for 1980–85 was 1600 million dollars, while 8500 million dollars has been allocated from 1986–90: 86% of the original fund was raised from taxes on the manufacture or import of certain chemicals and petroleum. The remaining 14% was financed from general tax revenues. For any particular clean-up, 90% may be payable from the fund with 10% by the State Government. If a responsible party still exists, they will be made to bear at least a proportion of the costs.

HAZARDOUS WASTE MANAGEMENT IN DEVELOPING COUNTRIES

Overview

Information on hazardous waste management in developing countries is not generally available. Not only do most developing countries lack the laws and governmental institutions to deal with increasing hazardous waste control problems, most lack the technological expertise even to assess the extent that hazardous waste problems exist.

Drawing from several sources, the ISWA working group found that problems in developing countries with hazardous wastes stem from several sources. Problems have arisen from hazardous products produced in developed countries and shipped to developing countries for use in such industries as agriculture, raw materials extractions, and metal-processing operations.

In some countries especially rich in natural resources long-standing problems exist with the handling of mining and processing waste. In other countries, with new industries that produce hazardous waste as a by-product, serious efforts must be undertaken if this waste is to be managed satisfactorily. Also, some hazardous waste from developed countries has been exported to developing countries that lack the technological capability to treat or dispose of these wastes safely, although no firm data on how pervasive this practice might be is available.

In addition to the ISWA working group, the problem has attacted the attention of several international organizations. The Organization for Economic Cooperation and Development (OECD), the World Health Organization (WHO), the United Nations Environmental Programme (UNEP), and the World Bank (WB), all have made significant contributions to the field.

Hazardous waste problems in Southern Africa

Throughout Southern Africa, according to the ISWA report prepared by the Southern African Working Group on Hazardous Wastes, a growing concern exists within both the private and public sectors over the build-up of hazardous waste. In Swaziland, technologically designed disposal sites are non-existent. In Lesotho, a serious problem exists due to the presence of hazardous waste in domestic refuse bins. Chemical poisonings of both children and domestic animals have occurred. In Harare (Zimbabwe), problem hazardous waste is produced by the following industries: metal processing operations, tanneries, oil refineries, asbestos manufacturing, and chicken farming.

This waste is disposed of in quarry landfill sites where hazardous waste is mixed without the benefit of records. After years of uncontrolled disposal practices, however, the government in Harare created the position of Noxious and Toxic Waste Officer to deal with the problem. A permit system was devised and hazardous waste is being identified. In Transkei, on the other hand, there are no controls at all, and chemical waste is being disposed of on urban waste dumps. The waste receives no cover, and waste material is often washed into rivers.

2. SUMMARY AND ANALYSIS

The main problem wastes identified in the report are organic chemicals, pesticides, phosphogypsum, and other inorganics. Coal waste is identified as hazardous at times because of its likelihood to ignite spontaneously. Waste from gold mine and asbestos mine operations pollutes both aquifers and the atmosphere, although this problem exists mainly in South Africa itself. Problem technologies are identified as those relating to destruction of organic chemicals such as PCBs and pesticides, co-disposal practices, deficient incineration practices, and waste management practices requiring water because of insufficient water quantities.

The report identifies six areas of institutional and practical concern. The first is the lack of a well-informed body of public opinion regarding hazardous waste. The second is that a quantitative data base on hazardous waste does not exist, inhibiting adequate control. Third, there is generally an absence of specific national legislation governing the transport and generation of waste.

Fourth, according to the report, it is known that waste is being moved from more developed states to lesser developed states where disposal costs are lower, but it is not known what quantities are being moved and controls are inadequate to deal with the problem. Fifth, although co-disposal is thought to be workable for some waste under some conditions, the practice is not well controlled. Finally, oil spillage from tankers at sea, as well as road tanker washings, are not officially controlled.

On the positive side, the report claims that recycling is widely practised in Southern Africa. Phosphogypsum is used as a soil conditioner, metallurgical slags are used as rail ballast, and pulverized fuel ash is used in oil fixation processes and as building material. Lead, calcium, and tin are recycled. Old tires are reused as tarring for road surfaces. Buffing dust, rubber cut-offs, and plastic wastes are used as materials for new products such as dustbins. Also, the recycling of mineral oils is widely practised.

The report concludes that a direct need exists to improve the transfer of applied technology from the more developed subregions of Southern Africa to the lesser developed subregions. While these technologies are known in certain centers in South Africa, ways must be found to transfer this information and the associated skills to where they are lacking. Throughout Southern Africa there is a direct relationship between the degree of industrial development and the quality of waste management. If the environment is to be protected, the report concludes, steps must be taken now throughout the region to prevent practices that create serious environmental degradation.

Hazardous waste problems in Asian and Pacific countries

Through the efforts of the Japanese member of the working group and Dr Hay Htun, Regional Director and Representative for Asia and the Pacific,

UNEP, an overview of problems in the Far East and the Pacific countries was put together. Most of the developing countries in the region, according to Dr Htun, lack trained manpower for the efficient management of hazardous wastes. In addition, existing sanitation and environmental laws are not sufficient to deal with the complex problems of hazardous waste management. Many governments, however, recognize the need to protect human health and the environment, and are in the process of considering legislation adequate to deal with hazardous waste problems.

As is true in most developed countries, many Asian and Pacific countries lack data on the source, quantity, and characteristics of hazardous waste. Throughout these countries information on where and how hazardous waste is disposed of is lacking, although there is general recognition that this waste is mixed with domestic waste and disposed of in this manner. Without reliable data and information, Asian and Pacific governments are finding it difficult to formulate appropriate policies and strategies.

With increasing industrial development, the quantity of hazardous waste is expected to increase, and authorities are beginning to formulate management policies and strategies that take into account cradle-to-grave management of waste, according to Dr Htun.

Of the various treatment and disposal methods viewed as appropriate for the region, Dr Htun says incineration and landfilling are the most commonly considered options. Problems with incineration focus on high capital and operating costs and air pollution considerations. With landfilling, the establishment of adequate methodologies for site selection are needed, as are criteria for designing, constructing and operating these facilities safely.

Related problems in other developing countries

What is true of Southern Africa also is true of other developing countries, according to a consultant's report prepared for the working group. The report points out that generally types of wastes vary in direct accordance with the extent and nature of industrialization in a country. In India, major generators are identified to be the power industry, and industries producing organic chemicals, pesticides, pharmaceuticals, dyes and pigments, non-ferrous metals, fertilizers and steel.

In Egypt, most notably the industrial center of Alexandria, the major generators are pulp and paper manufacturers, textile producers, oil and soap processors, refineries, inorganic chemical producers, tanneries, the power industry, and bottling and dairy industries. In Nigeria, hazardous waste is generated by mining, oil drilling and refining operations. In Brazil, there are 24 petrochemical factories in Cubato alone. In Sao Paulo, which has 40% of the country's industrial capacity, the major industries are food processing, non-metallic minerals, metallurgy, metal-working, and textiles.

2. SUMMARY AND ANALYSIS

Most of these hazardous waste producing industries, according to the report, are located not only in the larger cities but in the most populated sections of these cities. The major exceptions to this pattern are extractive industries located at the source of their raw materials. Hazardous waste in developing countries presents potentially greater problems than waste produced in comparable cities of the regulated industrial world, the report states.

Another finding of the consultant's report is that much of the industrial waste produced in developing countries contains hazardous materials that are discharged as untreated effluents directly into sewer drains, where drains exist. This has had negative impacts on the steams, rivers, lakes, seas and the oceans. Also, the report finds, storage sites often are inadequate, with access unrestricted.

In a 1983 report on practices in India, Sundaresan *et al.* point out, "Most industries have so far discharged semi-solid wastes, toxic or otherwise, into wastewater drains," and much solid and sludge waste is dumped on fallow land in the public domain. Other revelations are that by-product phosphogypsum is landfilled or lagooned in slurry form, at most factories the land available for disposal is "nearly exhausted," and serious ground water pollution occurs from overflowing lagoons and leaching dumps. Also in India, approximately 16.6 million tonnes of fly ash was produced at coal-fired power plants in 1980, and subsequent ground water pollution has resulted from the indiscriminate dumping of this highly toxic, metal-containing ash.

Finally, the consultant's report notes, many developing countries totally lack laws that regulate hazardous wastes. In other countries such as India and Egypt, where some environmental legislation exists, the laws are not specific enough, not to mention the non-availability of trained personnel to control hazardous wastes. Sudaresan points out, "The existing Air and Water Pollution Control Acts (in India) do not have provisions for dealing with the collection, treatment, and disposal of toxic and hazardous waste, for which legislative action is necessary."

Activities and concerns of international organizations

A document prepared for the OECD's Waste Management Policy Group, "Council Decision and Recommendation on Transfrontier Movement of Hazardous Waste," addresses how to prevent shipments of hazardous waste to those developing countries unable to treat and dispose of them safely. Specifically, it would require OECD member countries to:

- Monitor and control exports of hazardous waste to final destinations outside member country jurisdictions.

- Apply the same strict controls on transfrontier movements of hazardous waste involving non-member countries that apply to movements involving member countries.

- Obtain prior consent of non-member countries to receive hazardous waste shipments, and notify in advance any country through which the wastes may be shipped.

- Prohibit movements of hazardous waste to non-member countries unless the wastes are to be received at a treatment facility that meets standard environmental requirements.

A second OECD policy group document, "Some Background Information Concerning Hazardous Waste Management in Non-OECD Countries," addresses quantities, sources, and procedures with which to formulate solutions. In terms of quantities, this document references estimates that "third" countries generate perhaps 20 million tonnes per year of hazardous waste. Of this amount roughly 15 million tonnes are produced by countries belonging to the Council for Mutual Economic Assistance (CMEA), mostly Eastern European countries. Approximately 5 million tonnes are produced by developing economy countries.

Three primary sources of hazardous waste in developing countries are identified as:

- Wastes generated by foreign-owned, state-owned, or joint-venture firms.

- Wastes generated by small entrepreneurs, farmers and householders.

- Wastes imported from other, usually more developed, countries.

The report cites OECD data from development research projects and technology transfer programs indicating the waste quantities in all of these categories are likely to increase rapidly during the next 10 years.

The report emphasizes that developing countries lack the resources to deal effectively with any of the above three categories, and will not be able to do so in the identifiably near future. The countries, therefore, must institute some means of monitoring and controlling hazardous wastes without setting up major new bureaucracies that would require expenditure of already limited public funds.

Six action possibilities are discussed. The first is the establishment of an information exchange between governmental entities and the generator. Without actually regulating the waste, the government would promulgate a list of waste substances, as well as identify the enterprises from which they might be generated. This information would be shared with potential generators as a means to identify waste and as the first step in assurance that some hazardous waste management procedures are applied.

A second possibility is to develop, with industry representatives, codes of practice that the generator would agree to follow. Codes and licenses in developed countries with existing regulatory programs could be examined for cues as to how to achieve both low-cost and environmentally sound programs. Similar approaches also could be taken to develop codes of good practice for pretreatment techniques.

A third means of assuring the safe disposal of wastes is to utilize the original contractural agreement between the investor and the host country. This document could specify in advance that the investor dispose of any hazardous waste generated at facilities in the firm's home countries. The OECD report points out that some large multinational chemical firms already assume this burden.

The report adds that many developing economies are requiring environmental assessments as prerequisites to accepting new industry. Such countries include, according to G. K. Sammy, Ecuador, Indonesia, Kenya, Malaysia, Phillipines, Singapore, Sri Lanka, Tanzania, and Thailand.

Another major problem for developing countries is the lack of funding and technical capability to deal with hazardous waste emergencies. Where large corporations are operating their own facilities, they likewise can provide emergency reponse capability. In numerous cases, however, licensing agreements and turnkey operations have resulted in hazardous waste producing industries being wholly owned by local nationals. Examples are steel production in India and metal and mineral manufacturing in Korea. Some safeguards might be included in licensing agreement, but this also can increase purchasing and operating costs.

The report concludes that Environmental Impairment Liability (EIL) insurance would seem to be desirable, given the complex set of situations involving hazardous waste in developing countries. The problems, however, of obtaining such insurance, paying the premiums, and utilizing the funds for remedial programs are immense. Insurance, for example, is not a likely solution in the case of old abandoned sites. Here the three alternatives seem to be limited to the use of public funds, use of external emergency funds possibly from an international organization, or non-response to the problem.

The concern expressed in the OECD report is that, given the priorities for resources in developing countries, hazardous waste management issues are unlikely to be addressed. Any solution that may come about would be the result of actions by industry, other nations, or international agencies and organizations.

One possible solution, the report states, may reside in the creation of regional bodies to provide policy advisory services to the countries in the region. The regional body, according to the outline in the report, could suggest legislation, policy guidelines, and regulations, and could function administratively as the agency to receive notification of shipments and advise on negotiations with potential new industries. The proposed regional agency also could maintain an emergency response team to augment the needs of all countries in the region.

The pooling of talent and resources in this way, the report concludes, could serve to increase awareness and strengthen the political capacities of governments in a region to deal with both external and internal sources of hazardous waste.

The UNEP's Ad Hoc Working Group of Experts on the Environmentally Sound Management of Hazardous Wastes, at its meeting in Cairo, Egypt, December 4-9, 1985, approved what came to be called the "Cairo Guidelines and Principles for the Environmentally Sound Management of Hazardous Waste."

A consultant's document was considered at this meeting entitled, "Legislation on the Management of Hazardous Waste in Developing Countries." The report relied heavily on material already published by WHO in its journal, *International Digest on Health Legislation*. The group called for a more comprehensive review based upon questionnaire reports from developing countries supplemented by visits to the countries concerned.

This group likewise considered how assistance might be provided to developing countries to help implement approved UNEP guidelines for management of hazardous waste. The group recommended that a conference be convened on the technical aspects of hazardous waste management as this topic relates specifically to developing countries.

The working group expressed the view that industrialized countries might be urged to contribute financially towards funding such a conference and to assist in its preparation. It identified the WHO/UNEP/WB Technical Manual as a reference or background document for the conference. It also recommended that the UNEP prepare a comprehensive comparative review of the state of hazardous waste legislation in developing countries, data on the management of hazardous waste in developing countries, and information on the transfrontier movements of this waste.

The World Bank proposes to use the manual to determine acceptable hazardous waste disposal practices for specific projects, and to assist its staff in identifying suitable stand-alone projects that could meet hazardous waste disposal needs in developing countries. The WHO will use the manual to stimulate activities at the country level, such as providing workshops to train people to apply the information contained in the manual. ISWA was perceived as a possible contributor to the development and management of these workshops, as well as to the development and review of the manual itself.

Assessment

It is clear that the problems of hazardous waste management have indeed reached international proportions. While quantities may be smaller, and waste-producing industries fewer in number, the problem exists in developing countries as well as in developed countries. Further reasons for concern are that in many cases the problem is not being addressed by the developing country in which it is being produced or to which it is being transported. On the whole, major corporations have assumed responsibility for their own wastes. In many cases where the industries are small or locally owned, however, adequate responsibility for treatment and disposal has not been assumed.

The problem is compounded by a general lack of resource in developing countries to deal with the problem. Training, technical expertise, facility development, legislation, and the necessary governmental institutions, all in varying degrees have been found to be inadequate to the task of managing hazardous waste.

If the resources are to be provided they must be provided by other sources, according to those working in this area. An obvious partial source are the major waste-producing industries that locate in developing countries. Corporations, however, are by definition vested interests and cannot be expected to deal effectively with the full problem. Much of the assistance will have to come from developed countries, either directly or through international organizations such as those discussed in this report. To leave the problem unaddressed could have serious consequences for all concerned.

CONCLUDING ASSESSMENT

Overview of national control systems

So far, this report has focused on the status of control systems for hazardous waste management in the countries represented on the working group. The aim of this section is to present a rather more critical assessment of how the systems have worked in practice, examining problems encountered and lessons learned.

In a number of countries, development of legislation and introduction of an effective control system is still a priority. In other countries, much progress has been made over the last 10 years, although the list of "things done well" varies considerably from country to country. However, for a number of countries, some or all of the following statements can be made:

- Effective legislation is in place.

- Good collection and transfer systems have been established (notably in Denmark, Sweden and the FRG).

- Effective manifest system have been introduced to control waste transport.

- Most operating waste treatment and disposal facilities have been licensed.

- In some countries, well engineered and well managed facilities have been provided for hazardous waste treatment, incineration and controlled landfill.

Although much progress has been made, much remains to be done. Technology to control hazardous waste is relatively new and fast progress continues to be made; modern facilities using the latest technology soon become outdated. In addition, most countries have come to realise that legislation to control hazardous waste is not sufficient by itself, but much additional work is required to make the control system work in practice.

One of the strengths of this working group was the willingness of members to discuss mistakes that have been made and problems that have been identified alongside the success stories.

As discussed above, about 40 issues of concern were identified by the working group. These are presented for discussion under seven headings, which are discussed in turn in the following sub-sections:

- Problems of implementation and enforcement, including wastes currently escaping control.
- Providing, financing and controlling the use of recycling, treatment and disposal facilities.
- Encouraging waste avoidance, reduction and recycling.
- Obtaining adequate information.
- Cleaning up abandoned sites and avoiding similar problems in the future.
- Improving communication with the general public.
- Technical problem areas, subdivided into problem industries, problem technologies and problem wastes.

This chapter concludes with two further subsections. These are future trends and developments, and recommendations for future ISWA activities in hazardous waste management.

Problems in implementation and enforcement

It was noted previously that most countries with control systems in place are now facing difficulties with enforcement. This appears to apply even when there are strong central regulations. Data from most countries indicate that enforcement is better achieved at the local rather than the national level. Basic problems exist in very small communities, however, where resources and expertise tend to be lacking. This leads in turn to inconsistencies in standards and differing priorities for policing.

2. SUMMARY AND ANALYSIS

In some countries, central control over technical standards, for example for permitting landfill sites, is not strong, considerable discretion being given to the regional or local authorities. In such cases, inconsistencies in resources and expertise between authorities may lead to dramatic differences in, for example, the types of wastes allowed in landfills, in the standards of operation specified in site permits, and in the standards for policing and inspecting facilities and waste producers.

A number of recommendations can be made to improve the standard and consistency of implementation and enforcement. Better coordination between authorities is required. More resources need to be made available at the local level. More facilities need to be made available for the training of personnel in hazardous waste management. Central governments need to take more positive roles in improving both standards and consistency of enforcement. Recent initiatives have been taken in this field in such countries as The Netherlands and the UK.

This is an area where an international collaborative study would assist in sharing experiences and leading to improvements in national enforcement systems.

The general problem of obtaining adequate, accurate information on the quantities and types of hazardous wastes will be discussed more fully below. By definition, the information available pertains to waste within the control system. One may only speculate as to the absolute quantities of hazardous waste escaping control.

Hazardous waste may escape control in a number of ways. It is difficult to control the large number of producers who only generate small quantities of hazardous waste. Household hazardous waste generally finds its way into everyday domestic refuse. Some hazardous waste may be deliberately camouflaged, by mixture with normal commerical or industrial waste. Some waste is defined as being outside national control systems. Examples in a number of countries include waste oil used as fuel and mining waste.

This problem is discussed in all of the country reports where a basic system has been established that controls the greater waste arisings. Particular emphasis is to be found in the country reports on Sweden and Denmark. These countries have an established system for collection of waste at the municipal level. Their experiences have been, the better the collection service provided, the more hazardous waste that enters the system. In Denmark, it has been estimated from field trials that at least 30% of hazardous waste is presently escaping control. If this figure is reliable, then it has clear implications for other countries, where one might presume that collection is less efficient.

The only other estimate of the quantity of waste escaping control is that in The Netherlands, where it is put at about 10% of the total.

These problems can only be solved gradually. More resources are needed to improve:

- understanding of the nature and origins of these wastes;
- information and advice to waste generators;
- collection services; and
- supervision and enforcement.

The specific problem of household hazardous waste is causing concern in a number of countries. As discussed earlier, there has been much activity aimed at the separate collection of waste. A particularly interesting field test has recently been completed in Austria. This has led to the institution of a city-wide scheme in Vienna. The scheme is notable because it managed to maintain a high cooperation rate (50%) over a protracted period. Experience in Denmark suggests that both periodic publicity campaigns and the use of deposit schemes are necessary to maintain public participation.

Providing, financing and controlling the use of facilities

Legislation for the control of hazardous waste is almost impossible to enforce unless adequate facilities are available for the proper recycling, treatment or disposal of the wastes produced.

An example is provided by The Netherlands, where legislation was introduced in 1979, but adequate facilities to allow all hazardous waste to be handled within the country will not exist until 1988 when the new incinerator (1986) is joined by a new controlled landfill site. In the intervening period, The Netherlands has been, by necessity, one of the world's leading exporters of hazardous waste.

Particularly in those countries with recently introduced control systems, the lack of adequate facilities is cited as a major problem. However, as discussed below, difficulties in providing the right facilities at the right time exist in most countries.

In a number of countries, it is not the absence of facilities that presents a problem, but rather "unfair" competition between landfill and treatment facilities. Landfill tends to be less expensive than waste treatment, so that, given a free market, landfills tend to predominate, whereas treatment is often necessary for environmental reasons. In some countries, control of such competition is necessary. In other countries, such as in the USA, regulations specify that only treated wastes can be landfilled, thus eliminating landfilling as an option for more highly hazardous waste.

2. SUMMARY AND ANALYSIS 85

Thus, the problem can be alleviated by one or more of the following methods:

- constraining the choice of treatment or disposal options to be used for particular types of waste;
- controlling the standards for landfills so that they no longer represent a "cheap" alternative; or
- subsidising certain recycling or treatment facilities.

In many countries a combination of all three of these approaches is used. Most countries prohibit the use of certain options, primarily land disposal, for particular waste. In addition, strong national standards for the permitting of facilities may mean the effective prohibition of certain waste, although, as noted above, local variations may restrict the effectiveness of this in practice.

A general trend was noted earlier in many countries towards the concept of engineered landfills. An example is the USA, where recent regulations require, *inter alia*, the installation of a double liner system at the base of the site. The gradual tightening of such technical standards have the effect of increasing the cost of landfill disposal.

There is also an increasing trend towards restricting the use of land disposal for any waste other than residues from treatment. Similar (political) pressures are likely to restrict further the availability of sea dumping or sea incineration as legitimate options.

In most countries, some form of, generally indirect, government financial support is given to the provision of incineration and treatment facilities for hazardous waste. One argument is that, if industry is forced to move in one step from inadequate and cheap disposal to a well controlled but very high cost system, then the incentive for evading control will be too great.

In a number of cases, governments have provided incineration or treatment facilities with an initial subsidy, for example in the form of low interest rates, in the hope that the facility may thereafter be self-supporting. In the short term, additional treatment capacity will be required to deal with waste no longer acceptable at landfill sites. However, both high prices for waste treatment and other government policies will tend to encourage waste producers to generate less hazardous waste. Thus, in the long term, the quantities of waste for treatment may be expected to decline.

For such reasons, long-term financial support for some treatment facilities may be necessary in order to ensure their continued availability for the "hard core" of waste that cannot be eliminated or recycled. In the absence of economies of scale, the economic charge for treatment of the remaining waste could otherwise be such as to encourage illegal or less sound disposal practices.

One problem pointed to in a number of country reports is the lack of a national strategy or plan for the provision of hazardous waste facilities. Most countries require such planning on a local or a regional basis, but progress with producing such plans has been relatively slow.

In addition to the high capital cost of many facilities, notably incinerators, another reason why long-term planning for hazardous waste management is essential is the long lead-time required to build new facilities. Not only will construction take several years, but planning and permitting procedures must first be negotiated. In many countries, typical lead-times are in the range of 7 to 10 years.

The time required to permit facilities is a particular problem in the USA, where only 500 out of 5000 operating facilities have so far been fully permitted.

Local residents will generally oppose any hazardous waste facility. The strength of opposition may even be greater for treatment or recycling facilities than for landfill sites. Most countries have some procedure to arbitrate between local opposition to a facility and their "national interest". In the long term, the successful implementation of a coherent national policy for hazardous waste management requires a rational and publicly acceptable approach to siting facilities.

Planning procedures for recycling or advanced treatment facilities need to be simplified if proper hazardous waste management is to be ensured in the future.

The "state of the art" hazardous waste management technology for most organic wastes is incineration. However, incineration has the potential for causing air pollution, and the public is concerned about the possible creation of dioxins or other highly toxic by-products. In the long-term, there is the need for new technologies to replace incineration. A necessary part of planning for the future is thus to support research and development of advanced hazardous waste recycling and treatment technologies. Particular problem wastes are highlighted at the end of this chapter.

It is accepted in most countries that facilities need to be provided for the majority of hazardous wastes produced within the country. However, there may be small quantities of particular hazardous wastes requiring specialist treatment and for which the provision of facilities in every country is not feasible. For such waste, there will be a long-term need for transfrontier movement.

Particular types of waste and of facilities where this is presently the case include:

- PCBs and other highly chlorinated wastes requiring incineration;
- mercury for recycling;

- materials requiring long-term storage, for example in the salt mine in the FRG; and
- to a lesser extent, other wastes requiring "high technology" treatment or disposal.

In particular countries, such as The Netherlands, where there is a legitimate shortage of landfill capacity, then export for landfill is justifiable. In other cases, export for landfill will generally occur for economic reasons, in order to use a facility that is less expensive to operate either because controls are inadequate or because the authorities wish to exploit a source of foreign exchange revenue.

Because some transfrontier shipments of hazardous waste can be seen as long-term necessities, control procedures are required. A particular difficulty is in coordinating international definitions of hazardous waste, to ensure that such shipments are not used as a means of evading control.

It is relatively uncommon for any region within a country to be self-sufficient in its hazardous waste management. In most countries, there are no restrictions on hazardous waste movements across internal boundaries. A notable exception is the Federal Republic of Germany, where a permit must be obtained in advance for shipment between states. The long time required to obtain such permits is cited in the country report as a significant problem.

Encouraging waste avoidance, reduction and recycling

There is general consensus that the best means of hazardous waste management is to avoid producing a waste in the first place, to produce less of it or to produce a less hazardous waste. Failing this, the waste should be recycled or utilized for resource or energy recovery. Only then, should consideration be given to treatment or incineration.

Control systems for hazardous waste management are fairly recent in most countries. Efforts have thus far focused primarily on providing adequate facilities for advanced incineration and treatment of hazardous waste. There has not, in general, been sufficient emphasis on the encouragement of waste avoidance and recycling.

This is seen as a priority problem in a number of countries, including most of those cited previously as already providing either financial or technical support for waste avoidance or recycling. Such schemes are generally new and their early results have been mixed. They tend to suffer from a number of disadvantages:

- Financial support tends to be limited to new technologies rather than to technology transfer.
- Financial support of new technologies is often tied to the need for

publicity. This can be a severe constraint in, for example, the pharmaceutical industry where the need for confidentiality is important.

- The bureaucratic procedures involved in obtaining support are often so protracted that industry does not bother.

Schemes identified in Austria and the FRG appear to avoid most of these difficulties. Several countries already have, or are considering, legislation to ensure that waste avoidance is undertaken whenever practicable. The FRG has already amended the Waste Disposal Act to meet with this goal (November 1986). The proposed Austrian law would include the power to control the introduction of new products that would either be accompanied by the generation of hazardous waste or would themselves contribute to hazardous components in household waste. Similar sections are already to be found in Danish law.

There are at least two promising approaches to reducing the problem of hazardous waste:

- Redesign of consumer products to eliminate the hazardous components, for example a program in Japan aimed at reducing the quantity of mercury used in dry-cell batteries.
- Labeling products containing hazardous materials, both to inform the public and to provide an incentive for the manufacturer to develop substitutes.

Unfortunately policies aimed at controlling hazardous wastes may have the unintentional side-effect of discouraging recycling. The fact that hazardous waste destined for recycling is regarded as raw material rather than waste is viewed in a number of countries as a loop-hole in hazardous waste regulations needing to be eliminated. However, if such controls, particularly on transport, are rigorously applied, then it is feared they may pose a significant constraint on the recycling activity. A compromise in order to achieve both objectives needs to be found.

Improving information

In general, national data management systems are still in the early stages of evolution. One of the better systems is in the Federal Republic of Germany, but it is instructive that, to quote from the country report, "The Federal Government aspires to a better system for data collection in this field."

One possible deficiency in some data management systems regards their coverage of in-house treatment and disposal of waste. This is particularly so with those systems based on survey of manifests or trip tickets for waste transports, or of (commercial) treatment and disposal facilities.

Looking to the future, the country reports highlight some possible new directions:

- In several cases, the need for computerization is stressed.
- In the Danish report, it is pointed out that we have insufficient knowledge of the origins of hazardous waste. A good information system requires an understanding of the relation between waste generation and industrial production. Progress towards this end is made difficult but not necessarily impossible by problems of confidentiality.
- In Austria, work is being carried out on data collection from small hazardous waste generators, focusing on the types of industry and estimation techniques.

Improving information on hazardous waste is an area that requires both more effort and more research.

Abandoned sites

In those countries where abandoned hazardous waste disposal sites are perceived as major problems, present methods of locating sites and of cleaning them up are seen as requiring improvement. Some specific technical problems are highlighted below.

Alongside the problem of current abandoned sites, it is useful to examine policies designed to ensure that today's land disposal sites do not become the problem sites of tomorrow.

Herein lies the core of the difference of opinion between official policy in the UK and that in other countries regarding landfill management. In the UK, properly controlled co-disposal of hazardous with domestic waste is seen as a way of allowing hazardous materials to be gradually assimilated into the environment, whereas special chemical landfills are seen as long-term storage, imposing a potential problem in the future. It is in this context that co-disposal of asbestos has been highlighted as being a particular concern.

In most other countries, a very different view is taken. Co-disposal is viewed as dispersing chemicals, creating severe problems for clean-up should groundwater pollution problems be encountered in the future. The types of waste going to land disposal are restricted, and those that are allowed are concentrated in special sites where control over future development will be exercised.

This leads to an additional problem, to ensure that adequate funds are available for the long-term after care of waste disposal facilities. Such requirements are included in the conditions for site permits in a few countries, notably the USA. There are problems in implementing such conditions as the availability of long-term Environmental Impairment

Liability insurance is currently limited, due to the general excess of demand over supply in the world insurance markets.

Improving communications with the general public

An area requiring particular attention concerns the fear the general public has of anything to do with hazardous waste. It may not be possible to stop people thinking in terms of "NIMBY", but it is essential that people should have confidence in the safety of properly managed waste disposal facilities and that they should see them as essential parts of the national industrial infrastructure.

Attention needs to be paid both to improving communication with the public and to introducing innovative approaches to arbitration in cases of dispute.

There is a need for public education campaigns that explain the inevitable links between consumer goods and hazardous waste production, and also explain the national plan developed for the safe treatment or disposal of those wastes.

Public confidence in operating facilities is essential. Good relations with local communities can be promoted by visits, open days, and other public relations programs.

Active control by an independent authority must be seen to be effective. Confidence in the safety of both existing and proposed facilities could be improved through the increased use of risk assessment methods by those authorities. Such procedures have been developed in the UK, Denmark and The Netherlands. In Ontario, Canada, risk assessment has been used explicitly in the selection of a central treatment facility site and the results have formed the basis of public consultation at all stages of site selection.

Another idea is to provide public funds for local opposition groups, so that they can obtain professional assistance in preparing and presenting their ideas to a tribunal. This helps to eliminate the fear that the case for the facility is carried not on its merits but by the weight of the resources in its support. Such funds are made available in, for example, The Netherlands.

A major problem is that the benefits of a hazardous waste facility are national, while the disbenefits are local. Procedures should be considered for compensation to local residents, for example by provision of community facilities, in return for accepting a hazardous waste facility.

Technical problem areas

One of the particular questions addressed by members of the working group was the identification of technical problems in hazardous waste management. In this section, these problems are summarized mainly in tabular form, with cross-reference to the individual country reports.

2. SUMMARY AND ANALYSIS

Technical problems are differentiated into two categories:

- problem technologies, focusing on deficiencies in technologies for waste treatment and disposal; and

- problem wastes, focusing on those wastes for which existing technologies or control systems are considered to be lacking in some respect.

The problem technologies identified by the working group are summarized in Table 15. The most widely perceived is combustion of waste oil in small commercial or industrial boilers, which do not guarantee a high combustion efficiency. The impurity levels in such oils are often high, leading to considerable air pollution problems.

A more specific problem encountered in a number of countries is the contamination of waste oils with PCBs, where transformer "oil" has been, inadvertently or deliberately, mixed with waste oil for recycling.

As a result of these problems, a number of countries are in the process of either banning the burning of waste oil in small units or of introducing new and more stringent controls.

Among other problem technologies on which there was some consensus as to the need for further technical development were:

- air pollution control on high temperature incinerators;

- leachate treatment from landfill; and

- the identification and analysis of wastes on receipt at treatment and disposal sites.

Finally, it should be noted that any facility for hazardous waste treatment or disposal may become a "problem technology" if either management or regulatory controls are inadequate. A case in point is co-disposal of hazardous waste in municipal waste landfill sites. Even in those countries where this is viewed as a legitimate option, it is recognized that its acceptability depends entirely on proper management and adequate controls.

An alphabetical list of problem waste as identified during round table discussions in the working group is given in Table 16. Apart from oily wastes and household hazardous wastes, which have been discussed elsewhere, other wastes about which concern was most widespread were identified. These included contaminated soil; highly chlorinated wastes, particularly those containing PCBs; dioxin contaminated wastes; and mercury and other heavy metal contaminated wastes.

In all of these cases there are a number of particular problems. These include a general shortage of available treatment capacity, and an uneven spread of capacity so that international trade may be necessary. Other

TABLE 15
Problem Technologies

Technology		Nature of problem	Cross-reference to country reports (1)
General	Analysis	Inadequate analysis/identification of wastes on receipt.	Dk, F, Nl
Incineration	Air pollution control	Difficult to meet ever more stringent air quality standards particularly with high-chlorine content of waste.	A, Dk, SW, Nl
	Corrosion	Control of excessive corrosion.	A
	Feeding	Methods for feeding drums to incinerator unsatisfactory.	Dk
	Small boilers	Combustion of waste oil in small commercial and industrial boilers. Contamination, particularly with PCBs, can lead to severe air pollution problems.	A, Dk, FRG, J, Nl, Sw, UK, US
Mercury recycling		Occupational health and safety (control of mercury vapors).	F
Landfill	Leachate treatment	Need for improved methods.	Dk, F, A
	Drummed waste	Should be phased out as effective control is almost impossible.	UK
	Co-disposal	Unsatisfactory if controls are inadequate or not properly enforced.	SA, UK
	Groundwater clean-up	Inadequate technology for cleaning contaminated groundwater and soil.	US, FRG

(1) Index to Country Reports
A   Austria   Dk   Denmark   FRG   Federal Republic of Germany
F   France   I   Italy   J   Japan
Nl   Netherlands   Sw   Sweden   UK   UK
US   USA   SA   Southern Africa

TABLE 16
Problem Wastes

Waste	Nature of problem	Cross-reference to country reports (1)
Acid tars	Require specialist treatment; technology still under development.	F, UK
Asbestos	Best available option is landfill, problems with health and safety and with subsequent development of the site.	UK, Nl, F
Coal discards	Spontaneous ignition.	SA
Contaminated soils	Need better systems and facilities. Extreme shortage of treatment capacity.	Dk, FRG, US
Dioxin-contaminated waste	Public fears make some technically suitable facilities unavailable. Lack of capacity, particularly for incineration of solid waste.	A, Dk, FRG, US
Dusts	Safe handling.	Dk
Flyash from incinerators	Concern over trace contamination with dioxins	FRG
Heavy metal-contaminated waste	Most appropriate option often involves landfill. Long-term need for better and more economic recovery.	A, Dk, FRG
Highly chlorinated waste	Most appropriate option is incineration, but technical problems remain. General shortage of capacity, some countries have none. Political resistance to imports.	A, J. Sw
Household hazardous wastes	See text.	text
Laboratory and similar waste	Small packages of difficult wastes (see text).	F, Sw, UK
Mercury-contaminated waste	Technical and economic problems with recycling if mercury content is low. General shortage of recycling capacity, need for transfrontier shipment.	A, Dk, F, Nl, Sw, UK

TABLE 16 (continued)

Waste	Nature of problem	Cross-reference to country reports (1)
Mining waste	Enormous quantities of waste are currently outside the control system.	SA, UK
Oils	Control over burning contaminated oil in small boilers (see text)	See Table 15
Oils contaminated with PCBs	A particular manifestation of a more general problem. Defintion of a "PCB waste".	FRG, plus those in Table 15
Oily water (from tankers)	Collection and control of wash water from oil tankers (see text)	F, Nl, UK
Organic waste	1. Long-term need for innovative treatment methods 2. Extension of controls to cover more waste.	Dk US
PCBs	A more extreme example of the general problems of highly chlorinated waste (see above).	FRG, F, SA, Sw, UK
Phenols	Long-term need for better treatment methods.	Dk
Phosphogypsum	Large volume waste from phosphoric acid methods (fertilizer) manufacture.	SA (for recovery methods, see A, J)
Sludges	Problems in drying some sludges, also long-term need for better treatment methods.	A, Dk
Dilute sulfuric acid	Large quantities produced from titanium dioxide (white pigment) production. Major disposal route presently to sea, must end by 1989.	FRG

(1) For key to country reports, see Table 15.

problems include a political resistance to imports, and a long-term need for "better" treatment methods.

Future trends and developments

In this section, the working group has attempted to highlight both current problems and possible methods of tackling them. It is hoped that this will contribute to the future agenda for action on extending and improving national control systems.

The working group is seeking to identify a number of future directions, which it sees as important for hazardous waste control over the next 5–10 years. Among its observations, is the assumption that the nature of wastes will change. This will be due to:

- changes in the basic structure of producing industries;
- a trend towards smaller quantities of more difficult wastes (e.g. from the electronics industry);
- the advent of new technologies, new products, new industries and new wastes (e.g. from biotechnology);
- the emergence of different wastes of particular concern; and
- changes in definitions, including those brought about by improvements in analysis and by new toxicity tests.

A second observation is that controls will shift towards waste production. Possible trends here include:

- increasing government involvement in industrial production, necessitating new skills and expertise in government officials;
- considerations of waste avoidance and waste management will become more integrated into the mainstream of industrial management;
- integration of environmental controls on emissions to air, water and land, including regulation of recycling;
- incentives for recycling and waste reduction;
- encouragement for environmentally sound products; and
- increased waste minimization and treatment at source.

A third observation is that moves will continue away from land and sea disposal. Part of the rationale for this will be technical, but the driving force will be political. A consequence of the trend will be:

- increased development of new treatment technologies;

- introduction of incentives and other innovative procedures to ease the siting problem for new treatment facilities; and
- introduction of environmental standards for treatment plants.

A fourth observation is that trend towards the long-term care of waste management facilities will continue. Insurance or other means of ensuring the adequate aftercare of land disposal facilities will be developed to serve this need, although perhaps only in part.

A fifth observation is that hazardous wastes will become more of a problem in developing countries. As controls on waste production and waste management become more stringent, there could be a continuation of the present trend for the more polluting industries to move to developing countries. This should be accompanied by the gradual extension of controls over hazardous wastes both to other industrialized countries and to the newly industrialized countries in the developing world.

Recommendations for future ISWA activities in hazardous waste management

The working group will develop a new 3-year action plan, to be given final approval in 1987 for 1988–90. Among the directions currently under consideration are the following:

- organizing symposia;
- examining a general, more technical area such as research and development or waste treatment; and
- examining one specific technical or administrative area.

If the latter direction is followed, possibly study topics could include hazardous waste in developing countries, household hazardous waste, and waste minimization.

3

Hazardous Waste Management in AUSTRIA

GERHARD VOGEL and CHRISTOPH SCHARFF

Institut für Technologie und Warenwirtschafslehre der Wirtschaftsuniversität, Vienna, Austria

OVERVIEW

At present the amount of hazardous waste produced in Austria cannot be exactly determined. In 1985 approximate data pertaining to the proportion of hazardous waste produced by the industries and technologies in Austria were published for the first time.

Only a few large industries of international scale exist in Austria. They are: the Federal steel production, the chemical industry (fertilizer production, oil refineries), the paper industry, and the power plants. In the case of small industry, electroplating works, among others, are important hazardous waste generators.

Wastes which present problems are heavy metals, PCBs and other organic substances with high chlorine content, and sewage sludge.

In 1984 the Austrian Federal Institute for Health (Bundesinstitut für Gesundheitswesen) carried out the first study in communities, health institutes, and 10,000 plants throughout Austria to achieve a general overview of the amounts, types and places of generation of hazardous waste.

The conclusions of the data produced by this study were the basis for the Hazardous Waste Treatment Plan, stated in the Austrian Hazardous Waste

Act, to be collaboratively developed by the State, the provincial authorities, and the small and large industries. This plan was published in December, 1985, by the Federal Ministry for Health and Environmental Protection.

NATIONAL CONTROL SYSTEM

Summary of legislation

The legal framework for the disposal of hazardous waste consists of:

- The Hazardous Waste Act, including regulations for registration and disposal of hazardous waste, as well as "Durchführungsverordnungen" (implementation regulations) of the Federal Ministry for Health and Environmental Protection.
- The Waste Oil Act, with regulations for the treatment of waste oil suitable for energy recovery, oil refinement or recycling.
- The Special Waste Index lists the substances which qualify as special waste in Austria. Although not yet a complete listing, new editions are continually being released.
- The Hazardous Waste Index lists the substances which qualify as hazardous waste in Austria.
- Regulations of the Federal Ministry for Agriculture for the limitation of hazardous substances in sewage water and liquid wastes.
- Standards with regulations for the quality of liquid waste from photographic laboratories and printing works.
- The Emission Control Act, with regulations for the quality of fumes released from incineration facilities.
- The Act on Transport of Hazardous Materials, BGB1 209/1979.

Laws pertaining to special waste are determined by the Federal government, but the nine individual provinces are responsible for the implementation of these laws. In the case of household waste the laws are determined by the province and implemented by the communities. Therefore the Federal government is unable to regulate uniformly the treatment and disposal of waste throughout the country.

Currently new laws on waste management are being developed by several provinces. The new regulations will focus on the collection of hazardous waste and secondary raw materials.

For example, a 1986 ammendment to the Salzburg Waste Law states both the duty of separate collection of household hazardous waste by the communities and the mandatory participation in these collections by the public.

A new law on chemicals will actualize the existing regulations. Some of its most important items will be obligatory labeling and declaration and the possibility of prohibitions, which will allow the authorities to control the traffic with certain chemicals.

In 1985, a study group with the Chamber of Workers and Employees completed a draft of a Waste Avoidance Law. Its tasks will be both the reduction and the influence on the composition of waste. The Federal Minister of Health and Environmental Protection will be given extensive possibilities to control the introduction of new products. Debate on this draft in Parliament is expected for 1987.

History of hazardous waste management

Until now the disposal of hazardous waste was left to the companies that produced it. The lack of strict legislation allowed companies to carry out reckless disposal of hazardous waste leading to severe damage to the environment. Only since the development of the Hazardous Waste Act in 1983, was the treatment of hazardous waste submitted to defined legal regulation, although the appropriate facilities are still to be installed.

In spring 1985, the Federal Environmental Agency was founded. This agency, working under control of the Federal Ministry of Health and Environmental Protection, is a public institution providing information on issues of environmental protection. Among its tasks are expert reports, emission control and data collection.

In 1985, the first nationwide hazardous waste management plan was published. It contains information on special waste generation, the actual state and future needs of collection and treatment facilities, on the transport of hazardous waste, first assessments of the costs of special waste collection and disposal, reflections on organization and funding, and finally measures for waste avoidance, separate collection of secondary raw materials, household hazardous waste, and old and abandoned sites.

Licenses and manifest system

Licenses for collection and treatment of hazardous waste must be acquired from local authorities, who base their decisions on building regulations and the Trade Regulation Act. An additional license is also needed from the Landeshauptmann (governor).

The generator, collector, or disposer, of special waste is required to keep an annual record of the type, quantity, origin and disposal technique of these substances. This record must be presented for inspection upon request of the authorities concerned. This report is mandatory for owners of all kinds of special waste listed in the Special Waste Index.

All hazardous waste listed in the Hazardous Waste Index, regardless of quantity, must be registered with the provincial government by the generator within 3 months of its production. The same is valid for noticeable changes in the quantity of waste produced. The declaration must state, type, quantity, disposal technique, and code number according to the standard of the hazardous waste generated.

Collectors and disposers of hazardous waste must declare the quantity of waste and their disposal techniques every 3 months.

Transport of hazardous waste is carried out by private as well as public companies. Transport licenses are given by provincial authorities. The Austrian Standard Organization is currently developing specifications for standardized transportation containers. Generators of more than 200 kg of hazardous waste annually (more than 20 kg in the case of certain substances) must ensure that waste being transported is accompanied by a trip ticket ("Begleitschein") giving a description of the waste, code number, quantity, physical state, packaging as well as the generator's, collector's and disposer's names and locations. Custody transfer data for the waste must also be listed and certified. Six copies of the manifest are prepared of which each member of the chain including the provincial authorities receives a copy. The documents are matched by the authorities and shall serve as a basis for statistics.

Subsidies

The Environmental Protection Fund was founded simultaneously with the Hazardous Waste Act in 1983. On the basis of examined applications, the Environmental Protection Fund provides subsidies for environmental protection investments specifically in the areas of air pollution, noise, and collection, recycling and disposal of hazardous waste. Subsidies are available in the forms of financing of borrowing costs (6%), investment subsidies (up to 100%), non-repayable contributions, and low-interest loans.

The focus of the Environmental Protection Fund Subsidies Program is to replace polluting technologies with environmentally sound processes. Subsidies will also be granted for the development of hazardous waste disposal facilities. In addition the Environmental Protection Fund serves as a consulting service for small industries in the selection of appropriate technologies.

Responsibilities

All hazardous waste generators are responsible for the disposal of the hazardous waste so that human health is not endangered, no unacceptable annoyance occurs, no avoidable danger occurs to plants and animals and the environment is not avoidably polluted. Furthermore, waste generators are responsible so that neither fire nor explosion hazard is induced, no avoidable noise is produced, the presence and multiplication of pathogens, pests, and disease carrying animals is avoided and social order and safety are not endangered.

Only when the delivery of hazardous waste is accounted for in the form of a manifest or trip ticket can the responsibility for safe disposal be passed on to the hazardous waste collector or disposer.

The only major facility for the disposal of hazardous waste in Austria is the EBS (Entsorgungsbetriebe Simmering) incineration plant in Vienna.

Programs for dealing with old, closed, or abandoned sites

At present time an inventory is being taken of old, closed, or abandoned sites. In the Province of Styria a few hazardous waste sites had to be moved: one landfill contained metal plating sludge, in another case a later abandoned chemical plant illegally dumped hazardous waste. In addition, past practices have caused problems with boron, perchlorethylene, and TCE contamination of groundwater in certain areas. The estimated number of old or abandoned sites is 2000.

DEFINITION, SOURCES AND QUANTITIES OF HAZARDOUS WASTE

Hazardous wastes are defined as moveable properties to be disposed of for which registration and disposal is required for the sake of public interest. Hazardous wastes cannot be disposed of safely together with household waste, nor can it be disposed of together with household waste after being treated.

Solid waste is subdivided into two categories, refuse and special wastes. Refuse is further subdivided into household waste, bulky waste, industrial waste and street sweepings. Special waste is categorized as minimally or non-hazardous special waste and hazardous and extremely hazardous waste.

A study carried out by the Federal Health Institute compiled the first data on the quantities of special waste generated in Austria in 1983. This survey contains all kinds of waste not collected, treated or disposed of in public facilities. By definition the study refers to special wastes, yet only a few among them are to be considered hazardous. Table 1 shows that 1.3 million tonnes of hazardous waste were generated in 1983 and only approximately 400,000 tonnes were considered hazardous.

TABLE 1
Special Wastes Generated in Austria in 1983 (approx.)

2,000,000	tonnes	borrow material
1,700,000	tonnes	wash and rinse water contaminated with organic components
1,400,000	tonnes	sludge/pulp and paper industry
1,200,000	tonnes	sewage sludge
500,000	tonnes	sewage sludge contaminated with organic components
400,000	tonnes	sulfite liquor/cellulose industry
400,000	tonnes	wood/timber industry
300,000	tonnes	caustic sludge
260,000	tonnes	gypsum
250,000	tonnes	construction waste
200,000	tonnes	ammonia
200,000	tonnes	sewage water contaminated with metallic components
600,000	tonnes	11 other substances
3,600,000	tonnes	600 miscellaneous substances
13,000,000	tonnes	total (estim. 1983).

According to the study stated above, Table 2 presents the amounts of waste treated by various processes.

TABLE 2
Processes for Special Waste Treatment in Austria (1983)

2,120,000	tonnes	physicochemical treatment
280,000	tonnes	hazardous waste incineration
420,000	tonnes	hazardous waste landfill
530,000	tonnes	household waste incineration
490,000	tonnes	household waste landfill
14,000	tonnes	underground disposal

COLLECTION AND TRANSPORTATION SYSTEMS

International regulations on the transport of hazardous goods have been accepted by the national legislation for transport by road, railways, sea and air. Manifest and trip tickets systems exist simultaneously. Licenses for the transport of hazardous waste have to be acquired from the provincial authorities.

To date a number of private companies collect special waste, for example the "Gesellschaft für Umweltschutz und Beseitigung von Öl- und chemischen Abfällen" in Vienna. Similar facilities exist throughout the provinces. A list of hazardous waste collectors and disposers is available from the provincial authorities.

Waste oil collection systems

Waste oil is collected by means of standardized collection vehicles under private operation. Under the Austrian Waste Oil Act, all oils have been consigned to the EBS hazardous waste incineration plant for burning as a secondary energy source. But very little such oil is arriving; the private sector is keeping used oil to burn in order to satisfy its own energy needs. To date, a private recycling facility is planned.

Household hazardous waste collection

Separate collection of household special waste depends on the voluntary participation of the individual citizen in public collection campaigns. The purpose of separate collection is to lower emissions from landfill sites and incineration plants, to avoid groundwater contamination, and to facilitate composting.

Old drugs and medicines are voluntarily collected by pharmacies and deposited with pharmaceutical wholesale companies.

In 1985 a model experiment scheduled for 1 year was carried out in two districts of Vienna. All household hazardous waste may be deposited with mobile hazardous waste collection units. Status reports show a 50% reduction of hazardous components in the waste collected since this experiment began. As a result a network of 33 collection facilities will be spread over the entire city area.

Special collection systems have been introduced for hazardous waste from hospitals and waste oil collection from the food industry and restaurants. Collection and transport systems are locally organized, although current efforts are striving towards a nationally organized program.

TREATMENT, STORAGE AND DISPOSAL SYSTEMS

Incineration systems

EBS (Entsorgungsbetriebe Simmering) is the only incineration facility in Austria for disposal of special waste and sludge. A waste water treatment plant is combined with the facility. EBS is privately owned but is rented to the Municipality of Vienna.

EBS has four main incineration lines of which two are rotary kilns for the destruction of special and hazardous waste and two are fluidized bed units for dealing with sewage sludge. The kilns operate at 1000°C and have afterburners capable of 1200°C. Dwell time in the kiln is 30-45 minutes; exposure to the afterburner is for 10 seconds. Drummed wastes are fed directly to the furnace; the incinerated drums are recovered as scrap and sent to the steel industry for recycling.

The rated capacity of EBS to incinerate special waste is 100,000 tonnes per year. In 1985, some 52,700 tonnes were processed. In addition, EBS processed 58,000 tonnes solids per day of sewage sludge. The four incinerators produce 80 to 90 tonnes per day of fly ash; each rotary kiln creates 20-30 tonnes per day of bottom ash which is deposited in a government owned municipal waste landfill.

EBS supplies energy both for its own operations and for district heat. Steam drives two turbines to produce 5.5 MW of electricity of which 2.5 MW are used in the incinerator plant and 3 MW are used to power the waste water treatment operation.

Air pollution control represents a major challenge. At present, EBS is equipped with electrostatic precipitators but not yet with wet scrubbers. In addition, a method designed to dry the sewage sludge from 25% of solids to 17% of solids prior to incineration was not successful. Apparently methane is produced in the process causing an explosion hazard; all efforts to solve this problem have proven fruitless to date. The net result is that the overall throughput is reduced. The economic efficiency of the entire EBS operation is adversely affected as well.

In addition, a physicochemical operation for pretreatment in order to maximize the energy output of the waste to be incinerated is planned. The estimated cost of the needed improvement for air pollution control, sewage sludge drying and the pretreatment plant is about $22 million.

EBS accepts only small quantities of high chlorine or high sulfur waste; to deal with these, the afterburner temperature is raised to 1300°C, but considerable corrosion occurs. Most of the wastes of this type are shipped to Herfa-Neurode (Federal Republic of Germany). In addition, some wastes are sent to Basle, Switzerland, where industry maintains incinerators capable of dealing with dioxins and other highly chlorinated species. Within Austria, the chemical industry in the Linz area is reportedly planning to design and construct incinerators to handle their own waste, perhaps to be organized along lines similar to the GSB (Gesellschaft für Beseitigung von Sondermüll) in Bavaria.

In 1984, 43 incineration facilities were known, in most cases small scale in-house incinerators. This figure includes 38 facilities in hospitals.

The major current concern for incineration plants is a set of new air emission regulations in force as of 1 June, 1984 under the second Emissions Control Act which is administered by the Ministry of Construction and Technology. Under requirements of the Act new waste incineration plants must meet the following criteria:

Total particulates	50 mg/m³
Hydrochloric acid	100 mg/m³
Fluoride compounds	5 mg/m³
Sulfur dioxide	300 mg/m³
Zinc and lead	5 mg/m³
Arsenic	1 mg/m³
Chromium	1 mg/m³
Cadmium	0.1 mg/m³
Mercury	0.1 mg/m³

In 1984, 6 physicochemical treatment facilities were reported, which are mainly for in-house use only.

Landfills and disposal of waste in the land

There are approximately 500 sites where waste and non-hazardous special waste is deposited. Of these, according to local authorities, a maximum of 16 meet the basic safety requirements. In 1984 waste treatment in Austria was by landfilling (66.0%), composting (18.7%) and incineration (15.3%). There exists a small number of landfill sites for certain kinds of hazardous waste, e.g. one for metal plating sludge and one for oil-contaminated soil (under construction).

Underground disposal in mines or caves is not yet being used in Austria, although the construction of an underground disposal facility is mentioned as a planned activity in the Hazardous Waste Treatment Plan. Neither deep-well injection nor surface impoundment is practised in Austria. Co-disposal with refuse, construction waste or other non-hazardous materials only occurs in exceptional cases on specially licensed landfills. Landfarming or disposal of hazardous waste mixed with soil is not practised in Austria, nor is ocean disposal.

Energy recovery and recycling

Austria has four major incineration plants. Of the three which are for household waste, two recover energy for district heating, as does the EBS facility for hazardous waste. In 1985, 3.8 million tonnes of household, industrial, and hazardous wastes were burned for energy recovery in Austria. Two incinerators in Vienna have already been improved with very efficient wet scrubbers for air pollution control, a third one for the EBS is planned.

The efforts towards reduction of waste generation in the Austrian industry are numerous. The following processes serve as examples for environmentally sound low-waste technologies:

- The Lenzing magnesia bisulfite recovery process allows the recovery of MgO and SO_2 in the magnesia bisulfite pulping industry.
- The VEW Electro-Slag-Remelting-Process.
- Production of aluminum fluoride as a by-product of the production of phosphate fertilizer.
- Production of ammonium sulfate and sulfuric acid from phosphoric acid waste.
- Recovery of hydrochloric acid from the pickling process.
- Method for processing the caustic sludge, freed during the acid polishing of lead glass, to lead carbonate concentrates.
- Energy recovery in the catalytic cracking process.
- Incineration of tires in the cement industry.
- Process for recycling stretched thermoplastic materials in the manufacturing process.
- Powder metalurgical manufacturing process of parts made from sintered iron and sintered steel (avoidance of waste through direct shaping).
- Refining process of acid goudron (artificial tar).
- Silver recovery from photographic fixing baths and light sensitive materials.

In collecting secondary raw materials from households Austria only reached a below average rate. In 1985 just 9.7% of the total household waste was recycled (22.0 kg/resident and year). The recycling rate of secondary raw materials can only be increased by specially designed collection systems. For 1986, a 27% increment of the quantity is expected.

Programs to reduce waste generation

The most recent program is the Waste Management Plan of the City of Vienna. A vast concept for the avoidance, reduction, and recycling of waste will be realized by the city authorities. Secondary raw materials and hazardous components are to be collected separately at source in monomaterial collections systems with the maximum economically acceptable density of container placement in order to provide the highest convenience to the public.

An intensive information/education campaign will be carried out to motivate the public. Legal framework (refund systems and a new Waste Avoidance Law) shall support the effectiveness of this concept. The separation of industrial waste and the incineration of the combustible components providing a basic supply of energy for Vienna's district heating system are predicted. All waste incineration plants will be equipped with new air pollution control systems.

Imports and exports

Only the import and not the export of special waste must be licensed in Austria. Many companies export their waste directly, therefore there are no data available pertaining to the quantities exported. Known destinations of special waste exported from Austria are Herfa-Neurode in the Federal Republic of Germany as well as hazardous waste sites in the German Democratic Republic and in Hungary. Reportedly 30% of the hazardous waste generated in Austria is exported for disposal or treatment.

ASSESSMENT AND FUTURE DIRECTIONS

The scope of the Hazardous Waste Act does not cover production waste which is entirely recycled within an accountable time period in the production process of the same company. Therefore in-house waste is not affected.

In the case of waste oil the Hazardous Waste Act only applies when the oil is not suited for energy recovery, oil refinement, or recycling. Liquid production waste allowed to be dumped into the public waterways are also unaffected by the Hazardous Waste Act. Air pollution is regulated by the Emissions Control Act, although the efficiency of control is inconsistent.

In addition, the problem of controlling small generators remains to be solved.

The Hazardous Waste Act became effective on January 1, 1984. Since mid 1984 the mandatory reports and trip tickets must be directed to the provincial authorities. Nevertheless there is neither a current evaluation of the quantities and types of the declared hazardous waste nor any verification as to how accurately the reports are filled out. It can therefore be assumed that the official control system is not yet operating on an efficient level.

Often the disposal of special waste must be dealt with on a national basis. Unfortunately there is no national center that can keep track of data pertaining to all areas of waste management (central data processing unit) and provide an overview of the state of special waste disposal. The nine provinces already have agreed on the implementation of a nationwide hazardous waste information network and its technical feasibility has been tested successfully. Financial funding problems remain to be solved.

The two main problems of special waste treatment are the co-mixture of various types of hazardous waste in transport and the treatment of mixed wastes in general.

Future policies and developments in Austria will include the development of a hazardous waste treatment concept where some waste can be deposited in landfills without pre-treatment, some waste can be incinerated, some waste must be pre-treated and some waste can be recycled.

There will be implementation of refund systems for certain kinds of hazardous waste (especially household hazardous waste). Labels for hazardous products will be designed in order to ear-mark them both for the consumer and for separate collection and proper treatment.

Implementation of a nationwide hazardous waste information network is very important including completion of a nationwide inventory of old, closed, or abandoned sites. Appropriate technical equipment for the existing (hazardous) waste treatment facilities needs to be established.

Among the positive achievements is the founding of the Environmental Protection Fund which by financial support contributes to the development and distribution of low waste technologies and serves as a consulting service for small industries in the selection of appropriate technologies.

Separate collection systems for various kinds of household hazardous waste are currently being implemented in order to detoxify household waste and to facilitate its further treatment.

The Federal Waste Exchange in Linz provides a computerized information service on supply and demand of wastes and secondary raw materials.

The trip ticket system seems to work and to be useful, although it is not yet entirely implemented, enforced, and evaluated, mainly because of the lack of personnel. Therefore valuable information is still hidden.

The largest problem of waste disposal in Austria is created by the fact that the State determines the regulations and that the nine provinces are individually responsible for implementing them. It is therefore difficult to develop a legal system which could be uniformly enforced throughout the nation. In all cases which are not criminal offences, regulations of the Hazardous Waste Act are punishable by fines.

Further problems include the lack of public hazardous waste landfills in Austria and the inadequate equipment of the only incineration facility for hazardous waste.

The steps toward solving the problems of waste management should be dealt with in the following order of priorities: waste avoidance and waste reduction, recycling (including material and energy recovery), establishment of environmentally appropriate waste disposal of residual substances, and minimization of pollution and costs produced by waste management processes.

Therefore it will be necessary to influence product design and the selection of materials in production. Regulations will be issued with regard to processes and materials in production. Consumer guidance will be provided in the form of labels for recyclable or environmentally sound products and processes. Separate collections of secondary raw materials and hazardous substances will be organized and regulations and control systems for small generators will be set forth.

4

Hazardous Waste Management in DENMARK

KLAUS MULLER

National Agency of Environmental Protection, Copenhagen, Denmark

OVERVIEW

The terms "problem waste" or "hazardous waste" are used as a designation for those wastes which constitute certain risks. Chemical waste can constitute a risk to the human and biological environment, if it is not treated in an appropriate way; therefore all waste is subject to special legal regulations.

In Denmark, a law on oily and chemical waste was passed by the Danish Folketing in 1972; this law is a framework act empowering the Minister for the Environment to stipulate detailed rules governing the regulation of waste disposal. In this law chemical waste is not defined exactly, but described as that waste produced by the chemical or other related industries or waste arising from the application of products from those industries or other industries. A list of about 50 types of chemical waste is given, and no specific concentrations are stated. This list, which is subject to permanent extension, is divided into five main groups:

- animal and vegetable fats;
- organic compounds containing halogens;
- organic halogen-free compounds;
- inorganic compounds;
- other waste.

In order to facilitate the authorities' work, an index-system on different industries has been developed to identify typical waste streams produced in these branches. All technical investments, plants and production processes have to be registered and licensed by different authorities in order to obtain sufficient control of both the emissions and the fulfillment of the environmental regulations. There exists one central treatment plant and one central controlled landfill, both belonging to Kommunekemi.

In 1980 the National Agency for Environmental Protection (NAEP) began a study of 3115 landfills in order to evaluate their contamination with hazardous waste. Preliminary results show that about 900 old deposits contain chemical waste. The number of illegal landfills cannot be estimated.

Denmark has established a complete system and a set of procedures for the management of hazardous waste, including the following main components: legislation, packing/transport system, transfer station system and central treatment system (Kommunekemi). The three basic principles of this hazardous waste management system are: the obligation to give notification of the type and the amount of the produced waste to the municipal authorities; the obligation to deliver the waste to the municipal collection system and, the treatment of the waste at the central plant. The Danish NAEP will give a high priority to the following actions within the next years:

- cleaning up old, illegal or abandoned sites containing chemical waste;
- establishing a data-based system on types and quantities of hazardous waste from industries;
- establishing more service-minded collection approaches;
- promotion of recycling and cleaner technology activities.

NATIONAL CONTROL SYSTEM

Summary and legislation

The legal background for the Danish activities in the field of hazardous waste is given by Act No. 178 of May, 1972 on "Disposal of Oily and Chemical Waste". This act, which came into force prior to the introduction of the general "Environmental Protection Act" (October 1 1974) is a framework law. It has been supplemented by various notifications and circulars. It does not contain definite provisions for the management of hazardous waste. The objectives of the act are to prevent and combat pollution caused by oily or chemical waste and it covers the storage, transport and disposal of that waste. The Minister of the Environment may issue further regulations affecting both public and private undertakings as well as private citizens with regard to hazardous waste. Oily and chemical waste generated on board ships and the transportation of hazardous waste are subject to further regulations.

The history of hazardous waste management in Denmark can be told by the following milestones:

- 1971: Foundation of Kommunekemi A/S.
- May 24, 1972: Act No. 178 on "Disposal of Oily and Chemical Wastes".
- June 13, 1973: Law No. 372 on "Environmental Protection".
- 1974: Start of the construction work of the central treatment plant (Kommunekemi).
- November 1975: Official inauguration of the Kommunekemi plant.
- March 17, 1976: Notification on Chemical Waste.
- October 14, 1976: Circular on Chemical Waste.
- July 27, 1977: Notification on Waste Oil.
- September 1, 1977: Circular on Waste Oil.
- July 3, 1980: Notification on the Revision of the Notification on Chemical Wastes.
- 1982: Kommunekemi's second incinerator going on stream.
- June 8, 1983: Act Chemical Waste Deposits.
- July 1, 1983: Circular on Act on Chemical Waste Deposits.
- September 29, 1983: Circular on Review on Check up — and Reclaiming Operations according the Chemical Waste Deposits Act.
- April 24, 1984: Circular on Check-up — and Reclaiming Operations after Law on Chemical Waste Deposits.
- September 20, 1984: Notification on subsidies for Recycling and Cleaner Technologies.
- October 16, 1984: Act on Notification on Act on the Recycling and Reduction of Wastes.
- March 8, 1985: Law on Notification of the Environmental Protection Act.

Licenses and manifest systems

The Ministry of Environment is the highest administrative authority within the field of environmental protection, and the NAEP serves as its advisor and substitute. The administration of the Danish law system is organized after the decentralization principle, whereafter the local authorities in the municipalities or counties have to permit or license hazardous waste facilities according to the existing rules or laws, especially the Environmental Protection Act. These authorities are also responsible for the establishment of local collection arrangements for chemical waste.

Generators of oily or chemical waste have to inform the local authorities by using special information forms. The local authorities will review the information and give instructions regarding the delivery of the waste. This would normally imply the delivery to a transfer station for further transport to the central treatment plant. In case of transport by road, the waste must be accompanied by a security document called the Chemical Waste Card and a manifest describing the waste; if the waste is transported by rail only a manifest is necessary.

The administration of the Danish system is organized in a decentralized way. Therefore the municipal councils will take charge of the duties pursuant to the legislation on the protection of the environment, including the authorization of and permission for facilities, the administration of the waste collection system, exemptions and control. The regional and national authorities serve as appeal instances; complaints against decisions made by the NAEP will be negotiated by a special Environmental Appeal Board, which is an independent board not being subject to instructions issued by the Minister of the Environment.

Subsidies

There are no direct subsidies to enterprises for the treatment of hazardous waste, and each delivery of hazardous waste to Komunekemi is charged in accordance with a price list. On the other hand it must be mentioned, that special arrangements to facilitate the delivery of waste exist as well as municipal subsidies to Kommunekemi by granting interest-free loans or financing start-up operations.

Transport systems

There are two general transport situations: the transport of hazardous wastes from the producer to the transfer stations and the transport from those stations to the central treatment plant (Kommunekemi).

All transports are subject to special rules concerning vehicles, packaging, labeling, safety equipment and safety regulations; many of the requirements and rules are given on the Chemical Waste Cards which must accompany the waste during transportation. Furthermore, a special guide on emergency measures in the event of accidents is provided by the NAEP in order to ensure safe transport of hazardous waste.

The waste producer is responsible for the safe transportation of the waste from its place of origin to the transfer station, and he retains that responsibility even if the transport is carried out by a contractor.

However, some transfer stations do organize a pick-up service which means that the collection of oily and chemical waste is taken over by the transfer station by using the station's own contractor. This transport service is normally free of charge for the waste producer, and it is paid by municipal tax money.

Oily and chemical waste from private households is collected by means of a system of collecting points. According to the legislation there must be at least one such point in each of the 275 municipalities. From those collection points the waste is sent to the appropriate transfer station and then further on to Kommunekemi.

For the majority of the waste the transportation of the waste from the transfer stations to the central treatment plant (Kommunekemi) is carried out by rail. Drummed waste is transported by using open railway standard waggons, and bulk liquid waste is transported by Kommunekemi's special designed railroad containers.

Responsibilities

Principally, a generator of hazardous waste is responsible for the declaration, notification and delivery of the wastes. Even if he uses subcontractors for the transport, it is still his responsibility to treat the hazardous waste according to the rules.

A transporter of hazardous waste is responsible for complying with all rules concerning the transport of hazardous materials (safety equipment, packaging, labeling etc.). It must be mentioned, that the transporter is not responsible for the correct declaration of the waste on the Waste Card manifest. These documents are provided by the producer of the hazardous waste, and consequently it is not the transporter's responsibility.

The personnel at the transfer station have the function of advisors and instructors to industry with regard to delivery and transport of waste; at the same time it is the task of these personnel to control the conditions connected with the declaration type, quantity and characteristics of the waste delivered.

Programs for dealing with old, closed or abandoned sites

The NAEP started a project in 1980 in order to determine how many landfills represented a hazard because of chemical waste. Over 3000 deposits were checked: about 1300 are supposed not to contain any chemical waste, about 1300 are considered to be uncertain, while about 900 landfills contain chemical waste. Out of these 900 landfills about 110 deposits must be reclaimed immediately. The cost for this work is estimated to be at least 400 million kronen for the next 10 years.

DEFINITION OF HAZARDOUS WASTE

Existing legislation does not include a simple definition of the term chemical waste. Wastes are grouped into five categories:

- animal and vegetable fats,
- organic, halogen-containing compounds,
- organic, halogen-free compounds,
- inorganic compounds,
- other waste.

The purpose of this definition is to exclude any risk to the human and natural environment by inappropriate treatment. On the other hand it is desired to have legislation be as flexible as possible with regard to new technological developments and knowledge.

Industries producing waste according to the above mentioned definitions must report the types and quantities of waste to the local authorities, who give instructions to the industries and arrange a receiving system.

In theory, the definition covers all possible kinds of hazardous waste (except radioactive waste and waste from the production or use of explosives). The definition has a very high degree of flexibility, thus being able to cover future developments.

There will always be problems in defining waste as hazardous waste if they contain very low concentrations of the hazardous components. There will also be doubts and problems in those cases where new hazards, which cannot be compared with existing hazards, are discovered or where there are wastes, which by their quantity or way of treatment, constitute a risk for the environment.

There are specific regulations for all other types of waste not included in the regulations concerning hazardous waste such as household waste and office waste. The text of the notification on hazardous waste gives an explanation in those cases where the destinction between hazardous waste and other types of waste may be questionable.

SOURCES AND QUANTITIES OF HAZARDOUS WASTE

Table 1 presents information on types and quantities of chemical waste treated at Kommunekemi in 1985.

The waste quantities delivered to Kommunekemi have increased considerably from less than 20,000 tonnes in 1975 to over 90,000 tonnes in 1985.

There are data available on the waste quantities from various regions of Denmark. The figures show that the quantity of hazardous waste collected per inhabitant differs considerably from area to area. It is evident that the

TABLE 1
Chemical Waste Treated at Kommunekemi in 1985

Type of Waste	Quantity (tonnes)
Oily waste	30,670
Halogenated waste	1703
Solvents	5497
Organic chemical waste	32,070
Pesticides	1004
Inorganic waste	10,720
Miscellaneous	9534
Total	91,198

regions producing most hazardous waste per inhabitant are those where a pick-up service has been established.

The figures shown above represent those hazardous wastes delivered to Kommunekemi. The total amount of hazardous waste generated in Denmark is not known.

This is because the different communities can grant an exemption from the duty to deliver hazardous waste to Kommunekemi if environmentally sound treatment of the waste is ensured. Table 2 shows the quantities of waste for which exemptions were given in 1984.

TABLE 2
Exemptions from the Duty to Deliver Wastes to Kommunekemi in 1984

Destination	Quantity of chemical waste (tonnes)
Transport to foreign countries	11,600
Discharge to sewage systems	200
Deposits	900
Storing and transport (including future treatment within the industry)	5000
Burning/incineration	1100
Unknown activities	50
Other activities	4900
Total	23,750

It must be mentioned that exemptions often are given for certain waste quantities expected to be produced in the future. Also the last item in Table 2 (other activities) includes contractors who transport hazardous waste to other industries, and it also includes internal recycling activities.

Therefore it cannot be concluded that the sum of the delivered quantities and the quantities subject to the exemptions is the sum of all hazardous waste generated in Denmark. Furthermore, the quantity of those hazardous wastes treated in an illegal way is not known. It is known that a service-minded pick-up arrangement leads to an increase in those waste quantities delivered to Kommunekemi. A tightening of the control system shows the same effect.

COLLECTION AND TRANSPORTATION SYSTEMS

As mentioned before, all industries generating hazardous wastes have to notify local authorities who then will provide the industry with Waste Cards. These cards provide chemical and physical data and guidelines concerning packaging, management, labeling and transport. The Waste Cards are the first step of the control system.

The local authorities are also responsible for the establishment of a collection system. Each waste producer must complete a manifest (or multi-copy form) which gives information on the characteristics, components and process from which the waste was derived and the quantity and mode of delivery.

The supplier or producer of the hazardous waste delivers five copies of the form to the transfer station, where one copy is retained as a receipt for the delivery of the waste, and one is filed at the station. The remaining copies follow the transport of the waste from the transfer station to Kommunekemi, from where one copy is returned to the waste producer via the transfer station. This copy contains information on the laboratory examinations carried out at Kommunekemi together with calculations of the treatment charge.

There is at present no licensing system for transporters of hazardous waste, but of course all transporters have to fulfil the existing rules for the transport of dangerous goods. According to Danish legislation the producer of a waste is responsible for the waste, its transport and final treatment, and he has to supply the transporter with the necessary information and identification papers.

The transport of waste may be carried out both by private or public transporters. The delivery of waste to the transfer stations will in most cases be carried out by private transport. The transport of hazardous waste from the transfer stations to Kommunekemi will in most cases be carried out by the Danish railway system.

Waste oil collection systems

Since February 1, 1985, Kommunekemi has established a special collection arrangement in cooperation with the Federation of Mineral Oil Industries in order to increase the amount of waste oil delivered to Kommunekemi. The purpose is to prevent waste oil from being dumped or incinerated in environmentally unsuitable plants. Waste oil can at present be picked up on request or on regular schedules, and there is a payment for the delivery of waste oil graduated by the level of contamination in the oil.

Household hazardous waste collection

As already mentioned, a system of collection points, one in each municipality, has been established by the municipalities, according to legislation. This makes it possible for the public to deal easily with hazardous waste from private households. In Denmark the private person has the right but not the obligation to deliver oil and chemical waste to these collection points for disposal.

Use of this system of collection points is encouraged by local activities through advertising, pick-up schemes and mobile container services. Results show that it is possible to create a local awareness of the system resulting in a substantial increase in the amount of hazardous waste coming from households.

However, such campaigns must be repeated frequently in order to try to guide the attitude of the public in the direction of greater concern for the environment.

In an effort to make the system as easy as possible to use and especially to reduce the distances to the collection points special arrangements have been made in many areas between local municipalities and paintshops, so that the public can deliver old paints and contaminated solvents to these shops for disposal by the collection system.

An agreement has been made between the Danish Pharmaceutical Association and Kommunekemi where private households, doctors, veterinarians and hospitals can deliver leftover medicine to pharmacies, from where these wastes will be forwarded to Kommunekemi by means of the transfer station system.

Another collection agreement has been established to collect batteries, especially mercury batteries. At present used batteries can be delivered to battery purchasers, especially photographic shops, from where the batteries are returned to the original suppliers. From there they are sent abroad for recycling purposes or they are sent to Kommunekemi.

The Danish system has its background in national legislation, but the control, collection and transport system are the responsibility of the local authorities. Only the central transfer stations and the central treatment plant and the connected transportation system can be regarded as central elements; this conforms to the decentralization principle in the Danish environmental legislation.

TREATMENT, STORAGE AND DISPOSAL SYSTEMS

Storage systems

There are three possibilities for the storage of hazardous waste in Denmark: at 300 collection points (at least one in each of the 295 municipalities), at 21 transfer stations and at the central treatment plant prior to treatment (Kommunekemi).

The collection points are used for the receipt of oily and chemical waste from households, and generally they are placed near public plants such as municipal refuse dumps, storage yards, sewage treatment plants or similar installations in order to have personnel and control possibilities available. A collection station normally has a shed with a concrete floor and one side-wall of wire mask or grating in order to prevent the accidental accumulation of toxic or flammable gases. There will be drums for waste oils and other liquid wastes and at least two clamp-lid iron drums for other types of packaged waste. In most cases there will also be a special locker for highly toxic waste.

The transfer stations are typically provided with a railway track and a covered storage ramp for packaged non-flammable waste. Another separate covered section takes all packaged flammable waste. Besides this, a transfer station is fitted with tanks for the storage of liquid waste and sedimentation tanks for the pretreatment of oily wastes. There are also special designed railway tank wagons serving as receiving tanks for liquid waste bulk deliveries. An office building, a weighbridge, a shed for tools and other equipment plus a storage place for return-containers are other basic parts of a transfer station.

The Kommunekemi treatment plant

At the Kommunekemi central treatment plant there are the following main receipt and storage sections:

- railway tank wagon emptying section,
- drum emptying section,
- inorganic chemical waste section,

- waste oil section,
- receiving pit for other burnable waste,
- container site.

The railway tank wagon emptying section has vacuum tanks and corresponding pumps and pipe installations and heating systems. The drum emptying section consists of vacuum tanks with connected pumps, pipes and drum-opening installations.

Mixtures of waste oil and water are stored in a tank; solvent-containing liquid waste is also stored in tanks. All tanks are inert gas (nitrogen) blanketed, the nitrogen being supplied from a central tank.

For packaged solid waste, Kommunekemi is provided with two storage halls with a closed drainage system. One section of the central treatment plant is for treating inorganic chemical waste.

The various types of inorganic waste are mixed in three main groups:

- alkaline cyanide-containing waste,
- acidic chromate-containing waste,
- acidic iron-containing waste.

Additionally, waste containing hydrofluoric acid is treated separately, because this waste otherwise would attack ceramic.

The treatment is based on conventional stepwise detoxification processes (oxidation of cyanides by a hypochlorite solution, reduction of chromate by iron containing waste and neutralization by lime). The heavy metal content is precipitated as hydroxides and is removed by a filter press and deposed in a controlled landfill.

The incinerator plant burns liquid and solid organic chemical waste with a low sulfur and halogen content.

The incinerator plant accounts for Kommunekemi's largest treatment section. Solid, liquid as well as semi-solid organic waste can be incinerated here. The plant consists of a combined loading hall with two rotary kilns connected to secondary combustion chambers and steam boilers. The flue gas travels from the boiler through an electro-filter to the stack. For one kiln a NIRO gas absorber system is coupled between the boiler and the electro-filter. At the NIRO plant, suspended calcium hydroxide is sprayed into the system neutralizing the acidic components of the flue gas. The other kiln is likewise equipped with a flue gas cleaning unit which also neutralizes the acidic components of the flue gas. In the loading hall a crane for the feeding of bulk delivered waste has been installed, as well as a conveyor system with a drum lift for feeding packed solid waste.

Burner nozzles are mounted at the feed end of each of the rotary kilns for the injection of liquid and semi-solid waste as support fuel. Contaminated water, which is to be destroyed, is injected into the secondary combustion chamber. The energy value of the waste is utilized in a boiler for the production of steam of which 20% is used at the plant while the rest is sold to the city of Nyborg for the municipal district heating system. The quantity of heat produced is equal to an oil consumption of about 1.5 tonnes per hour.

The Kommunekemi faculty must be regarded as a public owned and operated facility, though the corporate structure of Kommunekemi is that of a share company owned by all Danish municipalities.

Landfills and the disposal of waste on the land

Filtercakes from the inorganic plant and slag, ashes and dust from the two incineration lines have to be disposed at Kommunekemi's controlled landfill, sited about 20 km from the treatment plant in a limestone area close by the coast of the Great Belt. The site covers an area of about 15 hectares and is subject to all control measurements according to the environmental regulations. Before being disposed at the landfill all deliveries must be controlled by Kommunekemi's laboratory, which is also responsible for the control of the surface water and groundwater as well as the precolate at the site.

Wastes which at present cannot be treated by Kommunekemi such as cyanide-containing hardening salts, mercury or PCB-containing solid wastes are deposited in West Germany in a salt mine.

Denmark has no underground disposal in mines or caves and underground injection of waste in deep wells is not applied in Denmark. Also surface impoundments such as pits, ponds and lagoons are not used in Denmark.

Co-disposal with refuse, construction waste or other non-hazardous materials is not allowed in Denmark. Land farms or disposal of waste mixed with soils is not carried out in Denmark. Ocean disposal is not used, because Denmark has ratified various international agreements which forbid this form of disposal. Denmark does not have any sea incineration system and there are no plans for establishing any.

Energy recovery and recycling

As mentioned above, Kommunekemi is producing steam in one of the incineration sections.

Kommunekemi has also established a waste oil plant with a capacity of about 25,000 tonnes per year in order to produce a fuel oil which will be used by Kommunekemi.

There are no other central recycling activities, but it must be mentioned that private industries are working in the field of recycling of hazardous waste, and there are also plans to re-establish a Nordic Waste Exchange. At present the most important recycling activity is the recycling of solvents. The NAEP is supporting research and demonstration projects for the recycling of hazardous waste, and it can be expected that there will be future activities especially in the field of recycling paint wastes and heavy or precious metal containing waste.

Programs to reduce waste generation

In 1984, the Danish Parliament passed the Recycling and Reduction of Wastes Act, effective October 1, 1984, which provides research, and demonstration and pilot projects within the field of low- and non-waste technology. Up to 100% of the total costs can be supported. This act has been passed in order to reduce the quantities and potential hazard of waste generated in Denmark. A number of demonstration projects within many industrial branches have been supported.

Imports and exports

It is difficult to estimate what percentage of hazardous waste is disposed of within national boundaries, and what percentage is transported to other countries for disposal. This is because of the different definitions of hazardous waste and because of the exemption system. However, exemptions from the duty to deliver wastes to Kommunekemi will only be given if an environmentally sound treatment of the waste can be ensured. Also, there are about 1000 tonnes per year (1985) of hazardous waste which cannot be treated in Denmark. At present, this waste has to be exported in order to ensure a controlled treatment.

ASSESSMENT AND FUTURE DIRECTIONS

The formal control system (notification, manifest/trip-ticket system) is estimated to function well in Denmark. There are efforts to improve the direct and local control measures by introducing a data base system on the industries' types and quantities of hazardous waste. Special emphasis must be paid to the enforcement of control measures, such as the enforcement of legal restriction of burning waste oil and clean-up activities of old, abandoned sites.

One current problem is the burning of waste oils in small plants. Another problem is the illegal dumping or transport of hazardous waste, often together with other (non-hazardous) wastes. A third problem not totally under control is the quantity of hazardous waste originating from households (such as batteries, paints, solvents etc.). Some improvements with regard to control can be made. Furthermore it would be ideal to have a precise knowledge of production processes and their corresponding types of waste.

Future policies will include a better control and understanding of production processes, enforcement of restrictions concerning waste oils, improvement and expansion of the existing treatment plant for chemical wastes and the promotion of recycling and cleaner technology.

In Denmark there is a good regional collection and transfer system. There is also good control of the industrial hazardous waste stream, cooperation between industry and the centralized management system and hazardous waste incineration has been successful.

Management of household hazardous waste needs to be improved and a small quantity generator management program needs to be developed. Complete knowledge of hazardous waste quantities has not been achieved.

Future needs include:

- an electronic data base for waste information,
- a better program on household hazardous waste management,
- a program to identify and reclaim old sites,
- a more agressive collection approach,
- promotion of recycling,
- restrictions on small incinerators.

Special attention needs to be paid to:

- dioxin-containing waste,
- illegal dumping of hazardous waste in the open sea,
- cleaning up of (undiscovered) abandoned landfills,
- groundwater contamination,
- heavy metal contamination problems with regard both to wastes,
- and recycling activities,
- information activities to industries and the public.

BIBLIOGRAPHY

Chemcontrol A/S: *The Danish System*. Chemcontrol A/S, Kopenhagen, 1980.

Kampmann, Jens: Benefits from and problems experienced with the Danish Hazardous Waste Management System. In: *2nd International Symposium on Operating European Centralized Hazardous (Chemical) Waste Management Facilities*. Odense, Denmark, Sept. 1984.

Palmark, Mogens: The Danish System. In: *2nd International Symposium on Operating European Centralized Hazardous (Chemical) Waste Management Facilities*. Odense. Denmark, Sept. 1984.

Warnøe, Kirsten: Situation in Denmark. Paper presented at *Scandinavian Seminar on Hazardous Waste*. Lindenborg Kro, Denmark, Sept. 27-28, 1984.

5

Hazardous Waste Management in THE FEDERAL REPUBLIC OF GERMANY

GERHARD SIERIG

Berliner Stadtreinigungsbetriebe, Berlin, FRG

OVERVIEW

There are many special disposal plants in the Federal Republic (about 120 facilities) with various possibilities for waste management — and special waste disposal is well organized in this country. However, some problems can still arise when new systems of reuse or disposal have to be found. For example sea disposal of dilute acid from titanium dioxide production must cease in 1989. The Federal Republic also stopped disposal of green salt in the sea in 1984.

Additional problems arise from the contaminated soils of abandoned industrial plants or from oil accidents, as the large volumes associated with such waste rapidly use up the capacity of disposal facilities and generate large transport problems. Other problems arise from PCB transformers and from PCB-contaminated waste oil. Also, the disposal of fly ash from waste incinerators partly contaminated with dioxins becomes more and more difficult.

Problems can also arise from existing technologies, which have been thought safe for a long time. An example is the recycling of waste oil, where authorities found PCB contaminants which had not been eliminated in the recycling process. Some problems and questions relating to this recently arose in the Federal Republic and have influenced the latest legislation for waste oil recycling.

There are many existing "state of the art" operations or facilities for special waste disposal.

Prohibited practices include uncontrolled dumping, mixing and diluting of special waste with water prior to discharging into surface water and burning without sufficient flue gas treatment.

Large facilities for treatment and burning of special waste on-site have been constructed by the major chemical producers. In addition, central treatment and burning plants under governmental responsibility are successfully serving the small and medium sized companies as are many smaller private companies specializing in certain specific types of special waste. More than 120 facilities can be listed for this country.

The Federal Government outlined the following priorities in October 1983, which have been implemented with the new Federal Waste Act in November 1986:

- more emphasis on avoiding special waste by new technologies and more recycling,
- more control of special waste treatment,
- less harmful effects on public health by treatment facilities,
- more control of dangerous residues which are going for off-site recycling and therefore not subject to waste disposal control.

NATIONAL CONTROL SYSTEM

Summary of legislation

The Waste Disposal Act of the Federal Republic passed in 1972 includes regulations for defining hazardous or special waste, guidelines for waste disposal, liability for disposal, regional planning for disposal, licensing of facilities, trip tickets, licensing for transport, waste import, irregularities.

The Federal Emission Control Act of 1929 includes reguations for operation of industrial facilities including waste disposal facilities, measurement of emissions, limits for emissions, irregularities, all with reference to air pollution and to noise.

The Hazardous Materials Transport Regulations outline identification systems for trucks carrying waste, and safety rules for this purpose.

The Environmental Statistic Act has regulations for collecting data on waste generation and waste disposal in the Federal Republic.

The Waste Oil Act with regulations for collection, recycling and burning of waste oil is now part of the new Waste Act of 1986.

The Act for Discipline in the Use of Water has special regulation against water pollution.

Licenses and manifest systems

As a result of uncontrolled transportation and dumping of special waste in the early 1970s in the FRG, special regulations have been set up since 1977 and 1978 for manifesting special waste and for licensing its transport with reference to Federal Waste Disposal Act.

Licenses for building and for operation of all waste disposal plants have to be acquired from local authorities on the basis of the Waste Disposal Act and the Act against Environmental Pollution. Most of the proposals or plans are necessarily discussed in public, and citizens may object when affected by the plan.

It is essential that waste generators manifest the type, amount and proposed method of disposal of special waste on trip tickets for local authorities, carriers and disposal plant operators. The generator, carrier and plant operator have to collect trip tickets in special books. Transport has to be performed in accordance with the Hazardous Materials Transportation Regulations. Solid waste and sludges are transported in closed or covered barrels or containers. Liquids are transported in barrels or vacuum trucks. The Federal Government has no responsibility for implementing or enforcing the requirements of the Waste Disposal Act. This authority is given to state governments (regional), mostly equipped with environmental protection agencies. Regional planning for special waste disposal in 5- or 10-year periods by state governments is required by law. Control of trip tickets and manifests is overseen by the responsibility of state governments; license for transportation and for building and operation of treatment and disposal plants must be acquired from local authorities.

Subsidies

Normally, subsidies are not given for special waste disposal, but they are sometimes given for studies and investigations in this field. Total waste management in the Federal Republic is performed under the so-called "polluter pays principle", which means the generator of the waste has to pay full cost for waste disposal. Subsidies will be given for waste oil treatment from a special fund as pointed out in the Waste Act, until 1989.

Responsibilities

Generators are responsible for analysis and for correct declaration of special waste on the trip ticket, and for determining the disposal plant authorized for treatment of the declared waste. Transporters are responsible for correct transportation ensuring that wastes are not mixed with other components, and for attention to Hazardous Materials Transport Regulations.

Treatment and disposal facilities are responsible for correct storage, treatment and disposal of the delivered waste. Laboratory checks are made to determine if the waste is accurately identified and whether the facility is fit and authorized for treatment. If not, recommendations for other suitable facilities are given.

Program for dealing with old, closed or abandoned sites

No special program is available at the moment in the FRG, nor is it feasible as different remedial actions have to be taken for different abandoned or closed sites. The state governments have knowledge of nearly all abandoned sites. Most of them are without major problems and only need a final cover. A few of them need special treatment, e.g. treatment of leachate, complete enclosure or excavation.

DEFINITION OF HAZARDOUS WASTE

Hazardous wastes in the FRG are termed "Special Waste". The purpose is to differentiate them from household garbage and construction waste. A second purpose is to list those wastes which have to be manifested on the trip ticket for transport. Data on special waste are collected as required by the Environmental Statistic Act. The existing catalogue for wastes in the FRG comprises hundreds of waste types divided into the following five groups:

- waste of plant or animal origin, e.g. from pulp production or from slaughter houses,
- waste of mineral origin,
- waste from chemical processes,
- nuclear waste (not controlled under the Waste Disposal Act),
- residential and commercial waste including sewage sludge and hospital waste.

Under the special regulations the Federal Government has listed those wastes which have to be classified as special wastes with manifest and trip ticket, e.g. salt from metal quenching, acids, infectious hospital waste and many others.

The definitions cover all known special wastes in the FRG. If new types of special waste arise, the catalogue can be amended. Not covered are single chemical components such as special waste (detailed composition) and toxicity. This would be too difficult to analyse. Special waste is waste that cannot be disposed of with household garbage because of type or amount, and therefore is in need of special treatment.

5. THE FEDERAL REPUBLIC OF GERMANY

SOURCES AND QUANTITIES OF HAZARDOUS WASTE

The FRG is a highly industrialized country where nearly all known technologies and processes are operated except perhaps some special types of mining and raw material production. Many kinds of special waste, therefore, are generated and listed in the catalogue. However, even with respect to manifests and trip tickets, it has not been possible to date to count all types of special waste, because data collection under the Waste Statistic Act is not comparable with the listing and collection of data on special waste, done by state governments from trip tickets. Data processing systems for general collection and calculation of all special waste data are not yet available in all states of the FRG. The Federal Government plans to improve the national system for data collection in this field by evaluating trip tickets.

The Federal Government estimated in 1983 the amount of special waste as listed in the catalogue at 4 million tonnes for 1980 and at 4.5 million tonnes for 1982. Waste generation from the most important industries, for 1980 is estimated as follows:

Industry	million tonnes
Chemical industry	2.70
Metal production	0.44
Metal works	0.51
Electrotechnics	0.15
Oil industry	0.07

Compared to about 30 million tonnes of household garbage and similar commercial waste in the FRG the amount of special waste is about 15%.

COLLECTION AND TRANSPORTATION SYSTEMS

Waste oil collection systems

Waste oil collection and treatment is now performed under the new Waste Act. Collection and treatment systems are mostly under private operation. Further experience will show whether regulations in the new Waste Act will optimize waste oil disposal compared to the old regulations in the abandoned Waste Oil Act, so far as licensed collectors are no longer obliged to collect any waste oil generated, from 200 liters or more, free of charge.

Household hazardous waste collection

Recently, local authorities started separate collection of special waste from households, such as used batteries, paint, solvents, medicines etc. The purpose is to lower emissions from landfill sites and from municipal incinerators. Usually, citizens deliver hazardous components to special collecting places, where containers and qualified staff are available, monthly or at other regular intervals. Collected material is then transported to special waste disposal facilities. These activities are quite expensive, and it is an open question whether they will spread throughout the FRG. Some kitchen waste from restaurants, hotels and hospitals is collected separately in bins or containers for use as food for pigs.

There is no nationwide centralized system in the FRG for special waste. Some Federal states have regionally centralized organizations. Sometimes, special waste disposal is locally organized. A nationwide licensed private company has recently developed from an association of some smaller companies, to collect waste oil.

TREATMENT, STORAGE AND DISPOSAL SYSTEMS

Storage systems

Storage of special waste takes place at first on-site, at the generator's facility, in bins, containers, tanks or bunkers. The materials are then transported to treatment or disposal plants. In some cases such as in Bavaria, special collection points have been built to reduce transport distances for generators or carriers, and to collect sufficient quantities of waste to make adequate loads. Sometimes such collection points are combined with pretreatment facilities for sedimentation. On-site at the treatment plants, storage systems for various types of waste have to be available, such as barrel storage, bunkers, tanks and containers.

Incineration systems

Land incineration systems in the FRG mostly use the rotating furnace. The main parts of such facilities are:

- bunker systems for fluid, sludges and solid materials, and tank systems for fluids, barrel and container storage,
- feeding systems for fluid, sludges and solid materials comprising jet burners, pipe feeding, elevator feeding with prechamber,
- rotating kiln furnace for burning up to 1000°C,
- discharging and cooling unit for residues,
- secondary burning chamber, which heats flue gas up to 1200°C,
- heat recovery system for flue gas cooling with steam production,
- flue gas cleaning system such as scrubbers and filters.

The plants generally have a weighing device, laboratory, supervising and administration area. One company, BASF, is operating six units. Others use one or two units. The capacity in the FRG for private and public incinerators for special waste is about 620,000 tonnes per year with 17 facilities. It has to be mentioned in this context, that most of the burning facilities include physical/chemical treatment equipment for detoxification, neutralization and precipitation. Moreover, 23 physical/chemical treatment plants for hazardous waste are available in the FRG.

It is estimated that about 30,000 tonnes per year of special waste from the FRG are burned on board ships in the North Sea, half of it highly chlorinated material, but no PCBs from transformers may be burned at sea. Those may only be burned in four existing land incinerators in the FRG. The Federal Government encourages all activities for recycling and reuse of highly chlorinated hydrocarbons. Today, only three companies in the FRG have licenses to deliver special waste for burning at sea.

Landfills and disposal of waste on the land

No waste disposal or special waste disposal is possible without landfills. About 22 special waste landfill sites are available in the FRG. Such landfills have to be located on sites with naturally impermeable soil or they must have artificial impermeable layers on the bottom. Contaminated water from the bottom has to be cleaned before discharge to a sewer. Generally, the groundwater in the nearby area is checked by drawing specimens from artificial collection wells. The types of waste received are: detoxified and dried hydroxide sludges from electroplating shops, sludges from grinding bricks and stones, slags from furnaces, ashes, detoxification sludge from special waste treatment plants and sewage sludge. There is only a small capacity remaining for landfill of hazardous waste. German State Ministers for Environmental Affairs, therefore, recently decided that each state of the FRG should provide sufficient disposal capacity both for burning and for land disposal.

One underground mine disposal facility in the FRG is Herfa Neurode, which is operated by Kali und Salz AG in Kassel. The facility is located in an abandoned salt mine. About 45,000 tonnes of special waste are disposed of here each year, e.g. metal quenching salt (cyanides), highly chlorinated residue from distilling hydrocarbons, pharmaceuticals, materials contaminated with heavy metals, catalysts and ashes from special waste burning facilities. Permission for disposal has to be obtained from the company and from the State Government of Hessen. Storage in the caves of the mine is manifested, so that special waste can be excavated later on for recycling.

Underground injections in deep wells is not performed in the FRG. Surface impoundments such as pits, ponds and lagoons are only sometimes used in the FRG for dredged sludge from the bottom of rivers and lakes.

Co-disposal with household garbage has occurred in the past on some landfill sites, e.g. for acid tar from waste oil recycling, oily sludge, contaminated soil, and detoxified sludge from electroplating. The percentage of special waste in such cases was about 5-10% of household garbage. The system is no longer recommended.

Land farms or disposal of hazardous waste mixed with soil is not usual in the FRG. Ocean disposal licenses are only available for diluted acid from titanium dioxide production until 1989. Sea dumping of green salt from the FRG was prohibited in 1984.

Energy recovery and recycling

The following systems for special waste recycling and reuse have been investigated or developed:

- recovery of waste salt from secondary aluminium melting and recovery of aluminium,
- pyrolysis of tires and plastics,
- production of animal food from kitchen waste,
- recovery of silver from photodeveloping effluents,
- recovery of heavy metals from electroplating effluents,
- oil recovery from waste oil by secondary refining,
- drying and burning of sewage sludge with waste charcoal with energy recovery,
- recycling of lead from car batteries,
- recovery of over-spray in painting facilities,
- reuse of zirconium waste,
- energy recovery from special waste burning by steam production for heating or electric power,
- energy recovery from waste oil burning,
- chemical and thermal destruction of highly chlorinated hydrocarbons.

Programs to reduce waste generation

No program to reduce waste generation is available at the moment but there are discussions of this subject in the FRG. In 1983, 19 million DM were given for investigations into reduction of special waste. Sponsored by the Federal Ministry for Investigation and Technology, a new system for chemical and thermal destruction of highly chlorinated hydrocarbons is under development using a stirred bed reactor system. The companies Nukem and HIM in Hessen are involved with these works. Also important are seminars on new recycling and reduction technology, and proceedings for industrial and public waste management held by the Federal Environmental Protection Agency. The Waste Act, including legislation and regulations to aid the change from waste disposal to waste reduction and recycling, has now been implemented.

Operation and ownership of facilities

Special waste disposal in the FRG is performed under the control of the Federal states. For the ownership and operation of facilities three kinds of organization are available: public, private and public–private cooperation. In Bavaria there are public organization formed by an association of various local authorities (Zweckverband Mittelfranken facility in Schwabach). Also in Bavaria, public–private cooperation exists in the ownership and operation of the facility in Ebenhausen (Bavarian Company for Special Waste Disposal). Both systems are compulsory and must be used by all generators and carriers. Large chemical companies mostly use their own disposal facilities, and state governments can oblige them to accept waste from other generators if facilities are appropriate.

In Hessen there exists a privately organized but publicly owned company with a facility in Biebesheim (HIM). In Nordrhein-Westfalen and in Niedersachsen there are many smaller privately operated facilities, mostly combined with collection companies. In Berlin (West) collection of special waste is done privately and disposal is performed publicly. Smaller private companies generally specialize in particular waste types. The on-site facilities of the large chemical companies have the advantage of a good knowledge of the generation and composition of the waste. Centralized publicly owned treatment and disposal companies are also useful, where transport distances are not too great. They provide an additional overview of the special waste situation in the Federal states.

Imports and exports

In 1982 special waste generation in the FRG was estimated to be about 4.5 million tonnes. About 40,000 tonnes were imported from Switzerland, Netherlands, France and Belgium, about 180,000 tonnes were exported — 140,000 tonnes into the GDR and 40,000 tonnes into EEC countries, Switzerland and Austria. That means about 1% was imported and some 4.5% exported.

ASSESSMENT AND FUTURE DIRECTIONS

Special waste recycling and the reuse of certain components as raw materials have to be enforced as well as the minimizing of special waste generation. Public money for investigation of the first two items can be acquired from Federal Government by industry. Special waste exchange markets have been operated for 10 years by German industry and can help to minimize or to avoid waste generation. Such organizations should intensify their efforts and should be recommended to each waste generator.

The control (trip ticket) system for transportation, and licensing of disposal plants is believed to work very well in the FRG. Within the last years, except for the problem of Seveso waste, no major violation in special waste management has been found in the FRG. Smaller amounts of special waste may still be mixed into non-hazardous garbage by smaller companies. Improvement of controls on-site at the disposal plants and more information for waste generators can help in this area as well as special separate collection for small generators of hazardous components similar to the collection systems for residential areas.

Data collection through Federal statistics should be improved, to obtain better knowledge about special waste types and amounts and about the flow of special waste in the FRG.

Waste disposal plants should be improved further to attain the higher technical standards developed in underground control for landfills and in flue gas cleaning for incinerators. More efforts are needed to inform the public about criteria for, and greater precautions in the building and operation of new disposal plants. People have to stop thinking in terms of the "NIMBY" principle (not in my backyard). The public's confidence in properly working waste disposal plants needs to be increased. New legislation outlining the economics of waste markets as well as new technologies for waste reduction has been drafted, and an improvement of the situation in special waste management is expected as a result. German building industry now offers safe landfill buildings for hazardous waste. New technologies are under investigation for immobilization of hazardous components and for abandoned sites.

The waste disposal program in Bavaria, with several collection points, is a good example of a system that works. The operation of waste disposal plants and the trip ticket system is done well in the FRG. Finding and licensing of sites for new treatment or disposal facilities is difficult and takes a long time due to obstruction by the public, but the number of modern, well-equipped treatment plants is increasing. Sometimes in the states of the FRG, special waste is generated, which can only be treated or disposed of in another state, and state boundaries have to be crossed. Acquiring licenses for such transboundary waste disposal is sometimes difficult, and the procedures should be improved. A better coordination among state environmental protection programs is needed. Mixing special waste components with other garbage still occurs and should be avoided in the future, to avoid malfunction of landfill sites and incinerators planned only to receive normal refuse.

The following issues need more attention:

- Procedures to avoid; avoidance of PCB-contamination of waste oil are needed (these will be improved by new Waste Oil Act).
- Better systems and facilities for treatment of contaminated soil from oil accidents and from abandoned sites are needed (mobile burning equipment for this purpose is needed).

- Sanitation of abandoned landfills bearing special waste needs to be improved.
- Discharge of salt from salt mines into rivers needs to be reduced.
- Improvement of salt recycling in secondary aluminum industry is necessary.
- Some definitions of hazardous waste need to be modified to end controversies as to how to dispose of fly ash from incinerators.

A high standard of living coincides with the high quality of the environment — in the air, soil and water. It can only be preserved when the amount of waste is reduced by new technology, and when waste disposal technology is improved further.

6

Hazardous Waste Management in FRANCE

JEAN-BERNARD LEROY

Compagnie Générale des Eaux, Paris, France

THE CURRENT STATE OF SPECIAL WASTES TREATMENT IN FRANCE

If the industrialization of France was primarily responsible for the proliferation of industrial waste—especially in the wake of World War II — a law dated July 15, 1975 signalled the actual advent of the age of rational treatment of those wastes. That law "makes it the obligation of each and every person generating or possessing wastes to ensure their disposal in such a way as to avoid adversely affecting human health and the environment".

The law initially:

- put an end to questionable practices such as unauthorized dumping;
- prompted the construction of treatment centers;
- prompted a fine-tuning of regulations, especially as regards landfills;
- prompted more elaborate techniques for collecting, transporting and storing wastes; and
- prompted manufacturers to make allowance for waste disposal in their operations and future projects. This resulted in the advent of the first test of "clean technologies", or more accurately "controlled-pollution technologies".

Several texts have since been appended to the above law. They govern classified establishments (1976), control of chemicals (1977), treatment of used oil (1979), industrial waste landfills (1980), incineration of industrial waste (1983), and the importation of dangerous toxic wastes (1983).

The general policy established by successive administrations has resulted in the rapid development of a number of industries. At the end of 1982, toxic waste treatment centers together had received FF 228.10 million in investments and showed earnings of FF 217.4 million, whereas the figures for landfills authorized to accept industrial wastes (known as class I landfills in France) amounted to FF 34 million and FF 41 million, respectively. In addition, the administration encouraged the setting up of a number of specialized bodies such as the ANRED (National Agency for Waste Recovery and Removal), which has come to play a major role in centralizing information and promoting new treatments and recovery techniques.

All techniques currently possible have been studied (dumping, physical-chemical treatment, incineration, upgrading, recycling and recovery). The regulatory activities of government authorities are limited to establishing discharge standards and seeing that professionals comply with them. It is left up to the laws of economics to establish a balance among the various possibilities. At present, the authorities are concentrating their efforts on bringing the flow of special wastes under control at the source, in order to reduce as much as possible the use of solutions economically unacceptable to the Commonweal. The problem of eliminating old dump sites, illegal but very real, remains. This is a priority task of the ANRED, which expects to come to grips with the problem in the next few years.

NATIONAL CONTROL SYSTEM

When the law of 1975 was enacted, the agencies then in charge of control were:

- the Department of Mines for unauthorized landfills and dumps;
- the Transportation Department for collection;
- Classified Establishments Agency for treatment centers;
- the River Authorities for liquid discharges in conjunction with the Hygiene and Occupational Safety Department, among others.

ANRED has since been compelled to play an increasingly greater role, and French authorities are in the process of drawing up a type of invoice system to keep track of special waste from source to final destination.

So as not to force conscientious manufacturers to bear the brunt of suddenly increased costs, a subsidy has been worked out for each River

Authority (1). The subsidy (originally about 40-60% depending on the authority) is supposed to decrease over a period of years. Allocation of subsidies is tied to sending waste to a treatment center or landfill approved for that particular waste.

Collecting and transporting waste are left up to the private sector, with one authority (Artois-Picardy) given powers to grant special approval. The two above activities, however, should be effected within the scope of prevailing regulations. The authorities control the whereabouts of waste by following up transport and reception records. Wastes are most often transported by truck owing to the relatively small quantities involved. There is rarely more than 100 tonnes of waste to transport at any one time. From time to time, however, rail transit is used.

DEFINING SPECIAL WASTES

History

Some time was required in France to define the concepts of "special waste" and "dangerous waste". The first to be involved were the utilities responsible for water quality. When they were first set up, the River Authorities based their fees on the water's content of oxidizable matter whose harmful effects are well characterized by biological oxygen demand (BOD) and chemical oxygen demand (COD) values.

It quickly became apparent, however, that some industrial discharges were extremely harmful. Such is the case with hexavalent chromium. It was necessary to introduce a further criterion based on the toxicity of the discharge as it affected precisely defined animalcules (in this case, daphnids). As a result, two families of distinctions were drawn up, one based on the source of waste materials and the other on their destination and, especially, dumping possibilities. Several nomenclatures have been established on the basis of the two lists, and an effort is currently under way to standardize them at the European level.

The current state of proposed definitions

Industrial waste in France can be separated into three major categories.

"Banal" waste is similar to household refuse (65% of the total), at least as regards the treatment processes involved. We should, however, refrain from lumping them together if only because of the large amounts involved, which can alter the nature of the disposal problem. A lumber mill or large wood processing plant located in a smaller urban area may have a great impact on local disposal. Nonetheless, the techniques to be implemented would be the same.

---

(1) There are six in France : Artois-Picardy, Rhin-Meuse, Seine-Normandy, Loire-Brittany, Rhone-Mediterranean-Corsica and Adour-Garonne.

"Special" wastes (32% of the total), which have characteristics specific to particular industrial activities, may contain toxic matter of varying proportions, especially in liquid form. Treating this type of waste may require special precautions, particularly if it is to be mixed with household refuse. Should this be the case, it must be done in exactly defined proportions and in a way that ensures thorough mixing.

"Toxic or dangerous" wastes (3% of the total with approximately one-quarter of this amount coming from the chemical industry and related activities) require specific treatments.

Dangerous wastes can be classified into three subcategories:

- organic wastes, especially those from the petrochemical and solvent industries, whose high chlorine content is a determining factor in their classification;
- liquid wastes containing toxic minerals such as acids, phenolated water with high COD, and used immersion baths for surface treatment;
- solid mineral wastes ranging from filtration soil to industrial waste salts and arsenic-containing substances.

The difference between the latter two categories is often a matter of one element's concentration, for example, chlorine. It may also be the result of a physical characteristic such as flash-point.

Nomenclature

In 1983, a complete nomenclature was drawn up by the Department of the Environment. The aims and means are detailed in a published document. The aims are improved definition of wastes, more effective control of disposal circuits, improved statistics, and creation of a common language.

In an effort to achieve this, the report sets up waste categories based on physical-chemical characteristics and treatment possibilities. Likewise, it identifies waste sources according to activity, the latter classified on the basis of a nomenclature drawn up by INSEE (the National Institute for Statistics and Economic Studies).

In addition to this very complete yet, for some users, slightly unwieldy nomenclature, there exist various lists. Among them is a list drawn up by the River Authorities. Others are intended simply to set rates, grant approval (wastes prohibited on certain sites), or announce a particular regulation. They should be considered complete and, as such, provide sufficient details. A recent example of a decree concerning the control of disposal circuits for nuisance-generating wastes is discussed later in this paper.

Difficulties encountered

Difficulties — the variety of the nomenclature provides a vivid illustration — are due in particular to the nature of various industries. Each of them is obliged to obey the laws of the market by improvising manufacturing processes and at times altering the products themselves. Furthermore, recent energy crises have prompted industry to recover all that could be saved, discharging only the truly unusable. Finally, the laws on waste disposal, and the resulting costs, have led to inclusion of this item in calculating costs and prices. Often revision of manufacturing processes have resulted.

Treatment centers are in a similar situation. They have to treat wastes of increasing difficulty, subject to increasingly severe discharge regulations and accordingly they must develop new treatment processes.

The current state of our knowledge is therefore already highly developed, but it is only a stopping-off point on a long route paralleling industrial progress itself. This trend in the advancement of our sciences is expected to continue.

THE ORIGIN OF SPECIAL WASTES

It is difficult to determine accurately the origin of special wastes for several reasons. Among them, the following can be cited:

- Detailed controls should begin at the entrance to treatment centers, and most landfill entrances do not have these controls. It is impossible, then, to gauge accurately the tonnage of waste treated. The tonnage of untreated waste is all the more impossible to evaluate.

- Identical waste can come from very different industries. Immersion baths for surface treatment, for example, may come from the car or metallurgy industries, the furniture industry, or from specialized workshops. On the other hand, one industry may generate several very different wastes. The car industry can be the source of the above immersion baths but also of foundry slurry, paint sediment, cutting oil, and other wastes.

- In-house treatment and the exchange or sale of wastes (such as waste-derived lime from acetylene plants) continuously alter both the amounts produced and the origins of wastes.

- The various government agencies have rightly shown more interest in what becomes of waste after it has been labeled as such by its producer, than in its processing before leaving the plant.

A report published in February, 1984 estimates that 32,000,000 tonnes of "banal" waste, which can be assimilated to household refuse, were produced. Likewise, 18,000,000 tonnes of "special" waste, characteristic of industrial activities, were produced. The latter figure contains 2,000,000 tonnes per year of "toxic or dangerous" wastes.

Based on various sources, their origins can be evaluated as follows:

- 40% from surfacing shops (integrated or not);
- 25% from the chemical or parachemical industry;
- 16% from engineering shops; and
- 7% from paint and varnish plants.

The remaining portion comes from the pharmaceutical industry, the food production and processing industries, as well as from hospitals and university facilities.

The report, prepared for the Department of the Environment, cites figures of 30-40% for waste generated by the chemical and parachemical industries. Smaller quantities of special wastes, whose total weight did not exceed 0.5% — or a maximum of 1% — of the total of all waste but which pose a high risk to public safety, come for the most part from farmers (pesticides and soiled wrappings), craftsmen and small industries, private parties, hospitals, road accidents, and closed factories.

The bulk of such waste originates in the research laboratories of universities and various industries. Certain harmless products, which can be reprocessed, may be included in this category. They include car batteries, silver and mercury batteries, and computer components containing precious metals.

Finally, the presence of a specialist is required when destroying 1000 tonnes per year. Between 20 and 50 tonnes per year at times require highly elaborate treatment and exceptional safety measures due to the special dangers they pose.

COLLECTION AND TRANSPORTATION

Before the advent of specific legislation on waste treatment, collection and transport firms also were responsible for waste disposal, meaning dumping.

Today, there are a wide variety of waste collection and transport firms grouped together in professional associations, such as the National Union of Liquid Waste Collectors and associations affiliated with broader organizations such as VANID (National Federation of Sewerage Unions). Many such companies also do industrial cleaning and work at contaminated sites. There is no official authority, with the exception of northern France.

In recent years, techniques have been markedly improved, especially in the area of liquid toxic wastes. They include increased use of multiple chamber cisterns and pumping systems, which have been better adapted to viscous and corrosive liquids. The increasingly technical nature of the processes has had the effect of instituting a type of natural selection and eliminating artisans who are no longer able to meet those new responsibilities.

Two technical practices could be substantially improved. They are sample taking at treatment centers and washing dumpsters and other receptacles.

Samples currently are taken on a very small scale. It is considered sufficient to mix a few cubic centimeters of liquid sampled several times during emptying, or a few hundred grams sampled at random. Due to the generally heterogenous nature of waste, a very serious problem is posed. As part of an EEC-decreed program, a study on improving practices is under way, with the final report scheduled to be submitted in June 1986.

Several centers are equipped with truck washing systems, but the drivers are hesitant to use them, especially, it would seem, because of the time lost. This type of situation may cause serious breakdowns, because, when mixed, a previous product may interfere with the treatment of the next one.

Two types of transport continue to pose special problems. The first concerns drums. Each shipment brings with it drums of varying contents. They may contain toxics, relatively homogenous substances, or a mixture of widely varying elements, even including iron plates or old shoes. Drum identification, which is always requested by the centers, is not always satisfactory. This entails costly verification processes.

Dangerous wastes in small quantities also are of concern. Experience has shown that extremely serious accidents could have occurred several times because those in possession of wastes, as well as collectors, are not always very aware of the dangers involved. Today, it is most often the treatment centers themselves that handle transport of this type.

Regulations governing the transport of dangerous wastes, and the control of disposal circuits for nuisance-generating wastes, are of very recent origin. They were the subject of a decree dated January 4, 1985, effective from July 1, 1985. The decree requires a loading slip for each waste. It is to accompany the waste from its origin to its final destination, and should be checked at every stage by someone in charge, including manufacturer, collector, transporter and disposer. A copy of the loading slip is returned to the manufacturer who thus possesses proof that a competent authority has taken charge of the waste. Each party is responsible for keeping a register showing all the operations performed on the waste until it reaches its final destination. The registers are available to the various Classified Installations Agencies within the several Regional Divisions of Industry and Research.

Conditions governing the exercise of dangerous waste transport will be stipulated in a further decree, as called for by the law dated July 15, 1975. Figure 1 shows the role played by each organization or body involved.

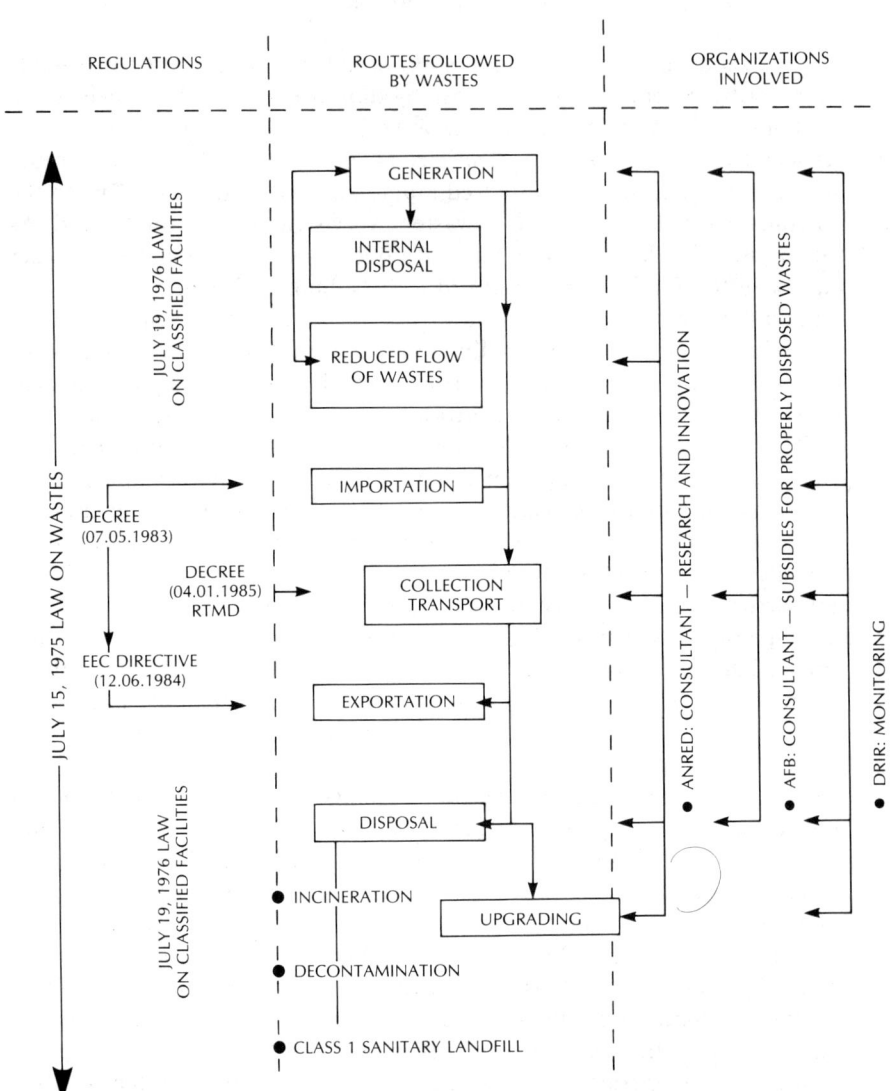

DRIR: Regional Agency for Industry and Research
AFB: River Authority
ANDRED: National Agency for Waste Recovery and Disposal

FIG. 1

6. FRANCE

French government authorities, treatment centers, responsible manufacturers and collectors, are all striving to establish and maintain an open policy in order to identify rapidly every waste product and its origin. Such a policy implies that the mixing of wastes will not be allowed, a very fortunate development, for treating mixed wastes always means running the risk of disastrous secondary reactions. Responsible collectors are now aware of this.

Of special concern under the decree is the practice of listing waste each time the generating quantity exceeds 0.1 tonne per month. Also, the invoice should show every operation including storing, regrouping or pretreatment.

The possibility of automating the processing of data contained in the invoice is explicity stated in the decree, which in return stipulates that the information will remain confidential.

The invoice contains six pages, the first of which identifies the generator. Besides the code number of the waste, it includes basic information on consistency and packaging. Also noted are the names of the collector-transporter and the recipient. The expected treatment also is mentioned.

The second page is to be used in the case of regrouping or pretreating the waste in order to identify the elements of the mixture. The other four pages are simply declarations regarding origin, transport and treatment, and the names of those in charge of each operation, without technical details. The declarations are intended for: the generator, the collector-transporter, the disposer, and the importer in the case of waste from other countries.

The slip drawn up by the generator of wastes should be returned to him by the final recepient. Should the generator not receive it within a period of 3 months, he must inform the authorities, who can in turn refer to the computerized invoice.

A decree dated July 5, 1983, regulates international transport of products shown on a list appended to the decree. In particular, a preliminary declaration is mandatory, with importation occurring, at the earliest 2 months after receipt of the document by government authorities. The same holds true for transit operations.

A form valid throughout Europe is scheduled to be drawn up and circulated in the coming years.

Approximately 20,000 tonnes of waste were imported into France during the first half of 1984. This represents 3% of the total amount treated annually, with two-thirds coming from West Germany and one quarter from Switzerland. Exports amount to about one-fifth of this quantity.

SPECIAL WASTE TREATMENT IN FRANCE

As of February 1984, wastes were treated as follows:

- Various treatment processes   5-600,000   tonnes per year
  (incineration, physical-chemical processes)

- Special landfills   500,000 tonnes per year
- Special treatment at the factory   300,000 tonnes per year
- Upgrading   200,000 tonnes per year
- Incineration at sea   11,000 tonnes per year
- Exported   10,000 tonnes per year

(Of which 4000 tonnes went to salt mines in West Germany)

Given the always approximate nature of such figures, we can nevertheless say that more than 80% of special wastes are treated correctly, i.e., according to methods and precautions that offer every guarantee for competent treatment.

Depots and transfer stations are monitored closely to avoid careless mixing of wastes, which are then impossible to identify and treat correctly. The length of time a waste is kept at such locations is reduced to a minimum as is the case in the storage facilities of well-run plants. In France, only harmless products are injected into the ground or kept in lagoons. Such practices are extremely rare. Finally, the mixing of household and industrial waste continues to be met with a great deal of scepticism. It can, of course, improve the hydrologic balance but at the risk of dissolving and diffusing toxic substances into the soil. It can only be authorized in watertight landfills but these choice locations are then taken over by household refuse. The method is, at any rate, a bad one.

In France, there are currently 13 landfills authorized to receive certain dangerous wastes explicitly mentioned on an official list. Regulations can be waived by local authorities in specifically defined cases. Some half-dozen other sites are now being studied for development. Dumping sites are subjected to a thorough geological study and must be naturally watertight or carefully made so. Two considerations are of great importance and are likely to occur at any time.

The first is that accepting special wastes and banal or household refuse at the same site may be judged financially and even technically (hydrologic balance) attractive, but it may severely reduce the lifetime of the landfill.

The second is that the treatment of leachate is not as yet full-fledged, i.e. in compliance with purification standards contained in legislation on water. Various techniques are being studied at present and some look to be very promising, but must be confirmed by experience.

In France, there are at present ten incinerators of widely varying capacity and variety. Three of these treat only liquid wastes. Seven can incinerate solids, liquids and pastes.

Total capacity of those two groups amounts to 50,000 tonnes per year and 150,000 tonnes per year, respectively. The largest among them (SANDOUVILLE, near Le Havre) treated about 65,000 tonnes in 1984, the smallest (SOBEGI, in southern France) 5000 tonnes.

At three of the centers, heat is recovered and sold to outside clients. Capacity in those centers has now reached saturation point. In addition, a few cement factories (five in 1984) have been authorized to receive certain special wastes, with about 20,000 tonnes being removed at present.

The following main technical problems remain to be solved:

- The treatment and monitoring of smoke emissions. In a number of these centers, the content level of certain elements in the waste (especially sulfur and chlorine) is limited so as to comply with legal standards. Continuous measuring devices are being studied — reliability is the greatest difficulty.

- Sufficiently regular steam production, in the case of heat recovery; and

- Corrosion phenomena.

Solutions to the latter two problems are relatively satisfactory depending on the centers. In these cases, the skill of the operator is important.

Five centers, with a total capacity of 250,000 tonnes per year, are now in service. The techniques used vary from one center to another. Some include processes for treating very special wastes (phenolated water, cuproammoniacal effluents). The most common processes are neutralization, chromium removal, cyanide removal, and sludge fixing.

Physical-chemical treatments closely depend on the nature of the waste to be disposed of, and by extension the manufacturing activity. Accordingly, the level of innovation in this area is high and the amount of laboratory research great. The following discharge-related problems remain to be solved:

- the composition of the sludge, which can vary according to the landfills expected to receive it,

- water discharge standards, whether it comes from sludge treatments or dewatering processes.

Standards governing the last point currently take into consideration COD phenol levels, and heavy metals. They also cover experimental leachates from fixed sludge. The main concerns of center laboratories are the disposal of new wastes or those until now considered untreatable, and compliance with future standards are expected to be more stringent.

Today there are ten centers in France for treating oil and solvents, with a total capacity of approximately 100,000 tonnes per year. Eighteen centers (a total of 50,000 tonnes per year) specialize in waste solvent recovery.

Three solutions are available in this field. The first is the recovery of waste oil and solvents for obtaining marketable products. This activity often presupposes a preliminary treatment for concentrating potentially useable elements, and is especially developed in the area of drainage oil. In France, regulations have organized the recovery system.

This second solution is concentration and treatment. This must be done either as final recovery, or the production of waste-derived fuels in liquid or paste form. Several processes are often found in this type of center owing to the wide variety of wastes.

Third is the destruction of wastes or the non-recoverable portion. The techniques used range from filter settling to "hot centrifugation".

Special mention should be made of cutting oils (which in the past were exclusively emulsions and have become more and more often solutions) if a costly and difficult incineration process is to be avoided due to the high water content and impurity level. Special techniques must be used such as evapo-incineration, which has recently been developed by the Limay center.

Acquiring exact information about in-plant treatment is obviously very difficult for two reasons: firstly, the nature of recovery processes changes with production techniques; and, secondly, the effort of individual enterprises in the field of waste recovery is closely related to economic conditions. The total capacity of these facilities has been estimated at 300,000 tonnes. Eight special landfills have been identified, especially in the fields of mechanical engineering, metallurgy and the chemical industry.

Small, highly specialized firms exist for the recovery and upgrading of wastes. Two of them recovery mercury, with some 30 shops working to recover silver from photographic effluents and dentist office wastes. Some half-dozen firms are engaged in the recovery of precious or strategic metals such as copper, chromium, nickel, and cadmium. One is devoted to treating computer debris in order to extract "bars of gold and silver".

Finally, 13 of the 22 regions of France have organized "waste exchanges" whose function is to bring producers into contact with potential users.

With this number of highly diversified firms, which are for that very reason well adapted to the terrain, France makes it possible for all manufacturers to comply with relevant legislation for which Government Authorities — in their role as guardian of the Commonweal — obviously are responsible.

FINANCIAL ASSISTANCE

In order to help manufacturers bear the burden of increased costs incurred in treating toxic wastes especially if the difference between the cost of treatment and that of using a sanitary landfill is compared, French Government Authorities have decided to allocate funds for transport and treatment. Funds are apportioned by the River Authorities, which are responsible for water quality and function like an insurance company. Initially set at almost 50%, of costs, financial aid theoretically decreases over the years before reaching a floor not to exceed 30%.

In practice, this type of subsidy is given once the waste has entered the center, with the River Authority then reimbursing the advanced sums to the treatment center. Subsidies are subjected to certain extremely stringent rules. The treatment must be approved for a given waste. A manufacturer may very well destroy, in compliance with requirements, a waste considered recoverable, but he loses any claim to a subsidy. Also, the center itself must be included on a list kept by each authority. A national approval procedure is being drawn up. When this becomes effective, controls by government engineers will guarantee a level of treatment and discharge in compliance with prevailing legislation.

The above method has made for a good start in the battle against pollution. Studies are being conducted, however, with the aim of transferring the financial burden from a water tax — as is currently the case through the agency of the River Authorities — to a direct tax on produced wastes. This type of tax would be collected for the government, which could in turn allocate a portion of it to semi-private organizations such as the National Agency for Waste Recovery and Study (ANRED).

As a rule, the Department of Mines is responsible for seeing to it that waste follows the prescribed routes and that class I landfills are properly maintained; the Classified Establishments Agency verifies that treatment centers are properly run; and the government's Technical Agencies monitor water pollution. The River Authorities and ANRED, by means of loans and subsidies, can promote research in and implementation of new treatment processes, and encourage development of methods for upgrading and recovering reuseable matter found in wastes.

It has been asserted that in 1985, more than 80% of toxic and dangerous wastes produced in France were more or less properly disposed of. The remaining 15-20% are either improperly disposed of, or are wastes for which no appropriate and economically practicable processes have been found.

ASSESSMENTS

General accomplishments and problems

Many aspects of hazardous waste management in France are working well. Processes for neutralization, and chromium and cyanide removal, are now well under control with discharge standards met without any particular difficulty. Incineration also is well understood, including that of solid and spadable wastes. Heat recovery is as high as 80%.

A new solidification process has just been patented, which affords a crushing strength of over 150 kg/cm^2, or almost that of concrete. Likewise, evapo-incineration processes (emulsions and oily solutions), and those for the treatment of liquid waste with extreme pH values and low fuel value (water, the mother of the chemical industry), are now operational in industry and authorized by the French authorities.

The same holds true for the recovery of hydrocarbons in solution with water treatment for removing phenols. Incineration of difficult-to-treat substances, namely chlorinated organics and PCBs, at high temperature (1200°C) with gas treatment by means of sudden cooling, is working well. Decontamination of PCB-contaminated receptacles using solvents, with PCB recovery, is achieving good results.

On the other hand, waste identification and sampling methods need to be improved. A study is currently under way under the auspices of the EEC. Another area of conern is the treatment of landfill leachates. Furthermore, there is a lack of approved landfills for hazardous wastes (known as class I wastes). A dozen or so are needed, especially in the west and southwest of France.

Draft legislation for levying a tax on materials and processes which generate hazardous wastes is under study.

As to collection managements, a body of regulations exists. It is expected to be given concrete form with a document of articles and conditions currently being drafted jointly by government authorities and industry professionals. It is expected to be ready by the end of the year.

Detailed trip tickets for monitoring the whereabouts of waste contain five forms, one of which is to be returned to the producer who is required to keep it at the disposal of the authorities (photocopy). Such forms will form the basis of future statistical surveys.

Import controls are deemed sufficient by the ministry. Attempts are being made to control export flows to KALI and SALZ and combustion at sea, which are the two principal destinations of waste. A European-wide form was drawn up July 22, 1985. Slightly complicated, the system is undergoing modification in accordance with French law with allowances made for the desires of the community.

The problem of collecting small waste amounts remains unresolved, and attempts are being made to integrate it into the regional system. In theory, service is free of charge for individuals, provided for a small fee to cottage industry, and provided for a fee to industry.

We should take note of the CORALIE-SARP system and a number of industry initiatives. The special case of farmers and the discharge of sacks containing toxic substances (pesticides) has not been resolved. French authorities now feel, however, that the juridical aspect is sound with further progress needed in application and monitoring. The removal and treatment of small amounts of refuse is a public service and, as such, should be seen to by local community organizations.

Specific problems

Acid tars are usually incinerated in cement factories following neutralization but can still be found in unauthorized dump sites. A special facility for incinerating acidic tars is located in Alsace. When they are found in abandoned landfill sites, they are treated on the spot by liming and mixing before being covered over. A number of specialized centers (SARP, SEDIBEX, GEREP) receive notable quantities of acidic tars.

Asbestos is discharged in sacks in class I landfills. A ministerial circular dated December 1984 provides details as to precautions to be taken during transport and utilization.

A circular dated August 1985 (technical instructions) provides information relative to regrouping on drumming of wastes, and transit and platform operation rules for drums. Such rules are related to safety and storage conditions. It should be noted that mixing waste products is considered to be treatment, and is monitored as such. Safety measures, storage and mixing conditions and the effect of facility size are all monitored closely.

The treatment of ship rinse water is called Marpol (marine pollution). By law, every larger port now provides ships with degasification facilities. All ships, however, do not use them. Current treatment yield for oil removal by means of settling is generally 50%, a figure that should be improved.

Some oil companies have their own facilities. In some ports (Fos) a portion of the recovered oil is sold as residue derived fuel with the sludge often sent to cement factories. A private concern (TANKER-Service, in Marseille) treats cleaning products in its own installations.

Hydrocarbon-water mixes are treated using more powerful means in some larger ports — Rouen, Brest, Nantes, Saint-Nazaire. Once again, not all ships respect the regulations.

Treatment capacities

PCBs are treated at Saint-Vulbas. Atochen at Saint Auban is pursuing research along those same lines and has taken out several patents. Condenser carcases can be washed and residual PCBs extracted from them (decontamination). There are two categories of wastes, depending on whether their PCB content is above or below 1%. Below 1%, treatment capacity is sufficient; above 1%, it is a bit close.

Mercury recovery facilities can be found in the Rhône-Alpes region (SNAM). The protection of personnel is a problem, which is as yet not completely resolved. The treatment technique itself is good. 75% of "flat" dry-cell batteries are now recovered — since used batteries can be sent free of charge through the mail. Most of the mercury from industrial catalyzers is now recycled by the industrial firms themselves.

The distribution of mercury in wastes is approximately as follows:

- 40 tonnes per year in the chlorine industry (internal recycling),
- 45 tonnes per year in 6200 tonnes of alkaline batteries,
- 3 tonnes per year in 22,000 tonnes of saline batteries,
- 7 tonnes per year in 20,000 tonnes of "flat" batteries,
- 15 tonnes per year from dental amalgams, and
- 10-11 tonnes per year from broken thermometers.

We should also add the mercury from phytosanitary products. Plant capacity (located in the Rhône-Alpes region) is just barely sufficient and should be increased. The main plants are: SNAM (now only treats Cd and nickel), and RAM (Hg + Ag and a few other metals).

Solvents are mostly recovered during the transition from the liquid to the steam phase (activated carbon recovery). Numerous industries have internal facilities for recycling solvents from printing, photogravure, and painting processes. Twenty companies are involved in solvent regeneration. The Ile de France has the oldest, hence the least advanced. They should be brought up to standard. The portion treated by these 20 companies is very small.

Approximately 350,000 tonnes per year of new pesticide products are produced each year (including poisons for crows, moles and so on). About 7000 tonnes per year are not used. They are dumped in highly diverse packagings in unauthorized sites. Treated presicide wastes are either incinerated (St-Vulbas) or dumped in salt mines. About 25% of the unused products are collected. The study is in its initial phase. Empty bags and other containers are burned at the edge of fields.

About 53 operations to decontaminate soil have been conducted over the 3 past years. They include a diagnosis, removal of the most highly toxic wastes, and on site treatment (solidification). An "oversight" office exists at ANRED (the appraisals and treatment prime contractor) for assistance in dealing with those problem sites.

Information and reports have been made more available (procedural reform, July 1983), with the purpose of better informing the public, comprised to a large extent of non-technicians. Information intended for the general public is obtained on a yearly basis from treatment installations. Prefects are empowered to grant information authorizations.

Prices have varied little in comparison to those contained in the SERVANT report. River Authority subsidies of 20-30% are available. No other aid exists, namely as regards investment outlays, except, in certain instances, when the recovery of deducted raw materials is involved.

Finally, the use of statistics will become a reality in France when the trip tickets are printed up. Statistics should be gathered on a quarterly basis in accordance with official requirements. Hence, there will not be any official statistics before an estimated 6 months to a year.

CONCLUSION

As satisfactory as the results may be, there is still room in France for improvement in the management of hazardous waste. The final conclusions of a February 1984 report on industrial waste disposal submitted to the Department of the Environment recommend the following:

- steps must actually be taken by manufacturers generating wastes to reduce the flow and harmfulness of those same wastes;
- strict monitoring of operating conditions in special waste treatment facilities must be ensured;
- disposal routes for wastes must be followed and maintained;
- study and research projects must be promoted;
- the collection and disposal of toxic and dangerous wastes of a not necessarily industrial nature must be better managed;
- information and training campaigns must be promoted.

For their part, treatment centers are striving to conceptualize and develop processes for wastes not yet properly treated. They likewise are working to improve existing treatment in order to comply with discharge standards that they feel will be more stringent in the future.

7

Hazardous Waste Management in ITALY

ALBERTO PIEPOLI

SEVESO, Lombardy Region, Milan, Italy

OVERVIEW

Italian industry was forced to face the problem of waste disposal with the EEC Directive and the Italian law of 1982. Before this, industrial waste was treated in the same way as urban waste. Only a few industries producing more dangerous waste took greater safety precautions. The new legislation on waste disposal created serious problems when it came into effect. A network of plants meeting the standards prescribed by the law or satisfied the demands of industry did not exist. The public generally opposes the installation of new waste disposal facilities, especially facilities such as incinerators, which produce environmental pollutants.

Industry does not always separate its by-products and some materials that can be re-used are mixed with hazardous ones. The possibility of retaining these by-products for re-use for other production cycles has not been given serious consideration. Many consider that strict legislation in this respect would stimulate renewed interest in the constructive use of by-products. For some years in Italy, a private company has been successfully managing the interchange of by-products between industries and also conducting research into manufacturing processes that produce more usable by-products.

Legislation on hazardous waste is recent, therefore waste disposal was previously uncontrolled. The old plants are now obliged to conform to the new legislation, whereas abandoned landfills need to be reclaimed, keeping environmental damage to the minimum.

Most success has been achieved by the system for collecting waste oil, the decontamination of areas polluted by toxic substances (e.g. Seveso, Arese) and development of methods for the dechlorination of surfaces polluted by PCB, PCDF, and PCDD.

The priorities for National action are:

- reducing waste at all levels of production, distribution and consumption,
- establishing an efficient method of waste control and censorship,
- creating an efficient network for the collection, treatment and disposal of waste,
- developing technologies for the recycling and detoxification of waste.

NATIONAL CONTROL SYSTEM

Summary of legislation

Italian legislation and regulation for hazardous waste is contained in:

- Law No. 691 August 1982, on waste oil,
- Law No. 915 September 1982, on toxic and harmful waste,
- Ministry of Industry regulation on waste oil, February 1984,
- Ministry of the Environment regulation on toxic and harmful waste, July 1984.

Laws 691 and 915 and related regulations resulted from the EEC Directive on waste oil, polychlodiphenyl and polychlorotriphenyl disposal and toxic and harmful waste management. Before these laws were passed, only a law regarding urban waste existed, together with some local regulations concerning sanitary landfill management.

Licenses and manifest systems

Licenses are needed for temporary storage, collection and transportation, treatment and landfill disposal. These licenses are only issued by the Regional Government if operators fulfill technical requirements and are financially covered.

There must be three copies of the trip ticket: one for the generator, one for the transporter and one for the treatment or disposal facility. This document must contain the transporter's essential references, the chemical/physical characteristics and quantity of the waste. All operators must keep a copy of this for at least 5 years. All the data on the trip ticket must be recorded in special registers.

The National Government has to establish the general criteria on treatment systems, the environmental limits, the cooperation between

Regional Governments, as well as collect statistical data and periodically give information on hazardous waste management to the EEC. Regional Governments have to decide the executive programs on waste treatment, confirm the site and the plants, and issue licenses for facilities. Local Governments must inspect facilities and verify the safety of waste management.

Subsidies

Financial subsidies or reduced costs are available to both private and public companies. Financial programs to this end are periodically prepared by the Regional Governments.

Transport systems

Regulations concerning the trip ticket must be adhered to. All carriers are obliged to use safe means of transport.

Responsibilities

The law prohibits the disposal of any kind of waste on the land, the production and disposal of hazardous waste that has not been registered and the transportation and disposal of waste without the necessary permit. The generator is responsible for the temporary storage of waste and the declaration of its composition. The carrier is responsible for the safe and separate transportation of each kind of waste.

Programs for dealing with old, closed or abandoned sites

The Regional Government is obliged to investigate old and abandoned sites, but at present there is no state financial aid for the reclamation of such sites. However, where abandoned sites have created health risks Regional Governments have acted independently and financed reclamation.

DEFINITION, SOURCES AND QUANTITIES OF HAZARDOUS WASTE

Italian legislation has three classifications of waste:

- urban waste (household garbage),
- special waste (industrial waste which by quantity or quality cannot be considered ordinary urban waste),
- hazardous waste (called toxic and harmful waste in Italy).

All waste which contains, or is polluted with specific chemical compounds above a determined level, or waste produced by listed industries is defined as hazardous. Laws regulate the concentration limits for each compound.

The fact that hazardous waste is defined either according to quality or origin doubles the amount of analytical control necessary. On the other hand, this system makes it possible to classify both the quantity and the quality of the waste produced.

The definition covers nearly all waste. It excludes compounds of amine waste and inorganic salts. Legislation covering this aspect is at present under examination.

An increasing number of legal restrictions exist in direct relationship to the harmfulness of the waste. The issue of licenses for the disposal of hazardous waste is directly controlled by Central Government Scientific Advisory Institutes.

Italian industries produce all kinds of waste but as yet there are no specific statistics available concerning the listed hazardous wastes or the sources of hazardous waste. The Ministry for the Environment financed a pilot study on the management of wastes in July, 1985 (quality, quantity and disposal facilities) in one region. This study is being carried out in a single area and the results should improve waste estimates for that area.

Information available regarding waste oil indicates that 130,000 tonnes per year of waste oil are generated, 500,000 tonnes per year are in emulsion form and 500 tonnes per year are dielectric oil (with PCB). About 50,000 tonnes of waste oil are not recycled; they are normally burnt by cement plants.

COLLECTION AND TRANSPORTATION SYSTEMS

Local authorities are responsible for the control of the collection and transportation of hazardous waste. There are two kinds of control: administrative in enforcing the law and issuing licenses, and the technical control of the safety of the means of transportation. All transported waste must be packaged and clearly labeled with the nature of its contents indicated.

The transportation and disposal of hazardous waste in Italy is carried out by private companies. Public services are limited to urban waste.

Waste oil collection system

According to EEC policy a public organization has been instituted to deal with the collection of waste oil. This is carried out by a network of private companies; the waste oil is ultimately sold to seven petrol industries where it is recycled. This activity has been in progress since May, 1984, and in the first 6 months approximately 30% of the total annual oil consumption was collected. In order to comply to this regulation everyone using more than 500 liters per year must record all consumption in chronological order.

Household hazardous waste collection systems

There is no program for the collection of household hazardous waste. At present only glass and paper are collected separately in some of the larger towns.

TREATMENT, STORAGE AND DISPOSAL SYSTEMS

Storage systems

The temporary storage of hazardous waste can only be authorized by the Regional Government, which regulates the maximum amount allowed (according to its toxicity), as well as the organization of its storage and security. The only nationally developed storage system is for waste oil. This organization covers the whole country. Seven central plants recycle the oil collected.

Incineration systems

Land incineration plants are divided into two groups:
- urban waste, special waste and hazardous waste with less than 2% of organic chlorine compound,
- hazardous waste with more than 2% of organic chlorine compound.

The requirements for these plants are excess oxygen, gas velocity of 10 m/s and 2 seconds residence time in the incinerator chamber.

For urban waste incineration the combustion temperature is 1050°C; for hazardous waste the temperature is 1200°C. An automatic system is also necessary to stop the loading of hazardous waste if the temperature decreases by 50°C. In 1986 three new incineration plants, approved for the processing of hazardous waste containing over 2% of chlorine, began operating. Two of the plants are built near chemical industries, and virtually their entire capacity is utilized by the surrounding factories. The third is privately owned by a company specializing in industrial waste.

Official data is not yet available on the potential of the plants or the quantity and quality of the waste treated.

No sea incineration system facility exists in Italy.

Landfills and disposal of waste on the land

According to waste classification there are three different kinds of landfills. The difference between these landfils lies in the safety criteria to be applied and the controls to which they are subjected. The main requirement is to protect groundwater, therefore regulations have been made concerning minimum soil permeability and geological conditions. There should also be a suitable network for the collection of gaseous and liquid percolates. Landfills for hazardous waste having a concentration 10 times greater than that established by law must have separate waterproof basins with individual systems for the collection of its percolates. The plant must be manned 24 hours per day by at least 2 people and an alarm system installed in case of accident.

Italy has no underground mines suitable for the storage of hazardous waste. Underground injections of hazardous waste is forbidden by law. Italy does not allow waste to be disposed of in surface impoundments such as pits, ponds and lagoons.

The law classifies urban waste with similar agricultural and industrial waste, construction waste, scrap motor vehicles and solidified treated sewage. This waste may be co-disposed with urban waste.

Only the waste water from farms or treated sewage may be dispersed over the land; and in any case must comply with health and hygiene regulations.

New legislation prohibits the disposal of waste into the sea. Until a few years ago one of the major Italian chemical industries disposed of its waste mud from the production of titanium dioxide into the Mediterranean.

Energy recovery and recycling

Recycling systems are mostly applied to materials of high value such as silver and heavy metals. An increased interest is being taken in the recycling of solvents by private companies. Other materials now being recycled are tires, coal ash and fly ash which are used by the cement industry.

Description of programs to reduce waste generation

At present the major undertaking in Italy to this end is the creation of a structure to deal with the hazardous waste currently being produced.

Ownership and operation of facilities

Existing facilities are mostly for urban waste, and are operated by both public and private enterprise. Hazardous waste facilities, however, are privately operated.

Imports and exports

Only an approximate idea can be given due to the lack of statistics. It is assumed that 35 million tonnes of industrial waste are generated every year of which 3500 tonnes per year are sent abroad for disposal.

ASSESSMENT AND FUTURE DIRECTIONS

Of major importance is that hazardous waste is collected and treated legally so that precise information can be collected on the quantity and quality of treatment. It is also important to program research into the recycling and reduced generation of waste.

Control systems are in their early days and therefore public criticism often slows down progress. It is hoped that once the initial stages have been overcome that control will be effective and precise.

Due to lack of adequate facilities and the growing costs of disposal both in Italy and abroad, much waste is disposed of with disregard to regulations. This does not apply to more harmful wastes which are mostly treated in specialized plants in other countries.

The control system being developed is extremely severe, but it is necessary to see how it works initially. Many of the regulations came into force in 1986.

Statistics drawn up on waste generation will permit adequate regulations to be made especially concerning toxicity limits. They will also provide a basis on which decisions can be made on where financial aid is necessary, in order to develop recycling and recuperation technologies. A method of convincing the general public to accept the presence of disposal plants must be taken into consideration if the problem of waste disposal is to be resolved in a short time in Italy.

Present legislation gives specific responsibilities to both regional and local government and private industry. It is very severe legislation and industry believes it is an imposition. The general lack of basic structures has increased costs and also illegal practices. Waste oil collection, however, works well. A minimal payment on delivery acts as an incentive and the network is continually growing.

The problem of household hazardous waste has not yet been approached. Neither a collection of processing network nor data on the production and management of this waste exist. There is not yet a program for the research and development of abandoned sites.

The law concerning the method of classifying toxic waste has been examined with extreme care in order to harmonize classifications with that of the EEC and to eliminate the costs of extra analysis and certification of waste. The government financing of the pilot study aims at revealing the objective difficulties of applying the new laws which might have been missed initially. Incentive is needed in the form of financial help for research into the recycling of waste, not only to help reduce waste but also to recycle highly toxic waste. For example much analysis is necessary to ascertain the absence of PCBs in waste oils. Special programs and financing are needed to develop abandoned sites on a national level, however, this problem is yet to be taken into consideration.

8

Hazardous Waste Management in JAPAN

SACHITO NAITO

Civil Engineering Department, Kanto Gakuin University, Tokyo, Japan

OVERVIEW

Problem wastes, industries and technologies

In Japan, wastes are categorized as municipal solid waste, which includes household waste, commercial and business waste, and industrial waste. The term "hazardous wastes" is not defined under the law, but this term applies to cinder, sludge, waste acid, waste alkali, slag and dust, found by extraction test to be above the prescribed hazard limit.

The following chemicals are subjected to hazard examination: alkyl mercury, mercury or its compounds, cadmium or its compounds, lead or its compounds, organic phosphorus compounds, hexavalent chromium compounds, arsenic or its compounds, cyanide, and PCBs.

Wastes other than the items named above, and also any of these which do not exceed the preset limit, are not subject to hazardous waste control.

The statutory regulations concerning hazardous wastes direct attention to the category named "industrial wastes," and impose restrictions on the hazardous wastes included in that category. For example, dioxin and mercury in municipal waste or waste oil in industrial wastes are sometimes called "problem waste" but are not regarded as hazardous waste.

It is essential to define the word "problem" in problem waste and problem industries, and then to specify which wastes and industries should be so listed.

Problem technologies are generally defined as unproven treatment technologies involving a problem related to environment pollution control measures or cost.

PCB wastes, for instance, have been stored for some time in Japan. As problems associated with storage have arisen, attempts have been made to dispose of PCBs by incineration and other methods. However, no proven disposal technology is available in Japan. Some problems were pointed out in respect of dioxins generated from municipal waste incinerators, but the Ministry of Health and Welfare has made an interim policy statement that dioxins do not pose a threat to the public as long as concentrations of dioxins in emissions from municipal waste incinerators do not exceed present levels.

Existing and closed or abandoned operations

Many instances of illegal industrial waste dumping under the Waste Disposal and Public Cleansing Law are exposed each year.

In 1980, 5456 cases of illegal disposal were reported from Police Stations throughout the country concerning water and air pollution control. Of these, 89% were from industrial solid wastes. Unfortunately, the cases of illegal disposals are increasing. In 1983, 5983 cases were reported.

The total amount of illegal disposal of industrial solid waste, such as construction and demolition waste, was about 270,000 tonnes per year in 1980, and 330,000 tonnes per year in 1983. However, these materials are not hazardous waste under the regulatory law.

The primary reason for illegal dumping has been the difficulty experienced by individual entrepreneurs in finding available disposal sites.

Many additional public disposal sites, however, have been made available to them. In April 1981, the total number of such public disposal sites throughout Japan reached 93.

Successful practices, priorities

Soil pollution by hexavalent chromium due to land reclamation with chromite-containing slag (1915–1973) posed a problem, but clean-up of the contaminated soil was carried out successfully by using the reducing agent, ferrous sulfate.

At a mercury refining factory, mercury has been recovered from mercury-containing waste, and treated. Disposed dry cells in municipal waste aroused concern as a source of dangerous mercury pollution, and this factory is being expanded to accept and process such dry cells.

Japan likewise has achieved success concerning a related mercury problem — how to maintain the inorganic mercury concentration at the WHO recommendation value (0.015 mg/nm^3) in emission gas incinerators. The Tokyo Metropolitan Cleansing Bureau reported success in July, 1985, for removal of 90% of emissions of inorganic mercury. This method utilized a liquid, which contained a liquid-form resin plus copper salt or manganese salt, in addition to the caustic soda spray system used to eliminate hydrochloric gas.

The Japanese government continues to investigate the state-of-the-art of hazardous waste treatment and disposal, and takes steps to detect illegal dumping sites. If illegal dumping sites are found, the authorities in charge give administrative guidance to bring about the necessary correcting action under the Waste Disposal and Public Cleansing Law and associated regulations. Also, they assist in developing appropriate technology related to treatment and disposal of hazardous wastes, and to demonstrate the effectiveness of such technology.

NATIONAL CONTROL SYSTEM

Summary of legislation

The Japanese Waste Disposal and Public Cleansing Law (Number 37) was enacted on December 25, 1970. It consists of five major chapters: General Regulations (Articles 1 to 5); Municipal Wastes (Articles 6 to 9); Industrial Wastes (Articles 10 to 15); Miscellaneous Regulations (Articles 16 to 24-2); and Penal Regulations (Articles 25 to 30).

Article 1 of the law states that it is enacted for the purpose of preserving the living environment and improving public health through appropriate disposal of wastes and conservation of a clean environment.

Article 2 defines "wastes" as refuse, bulky refuse, ashes, sludge, human excreta, waste oil, waste acid and alkali, carcasses, and other filthy and unnecessary matters, which are in solid or liquid form (excluding radioactive wastes or wastes polluted by radioactivity). Under the law "municipal wastes" refer to wastes other than industrial wastes.

"Industrial wastes" are those wastes discharged in connection with trade and industrial activities. They are as follows:

1. Ash
2. Sludge
3. Waste oil
4. Waste acid
5. Waste alkali
6. Waste plastic
7. Waste paper
8. Waste wood
9. Waste textile
10. Residue of animals and plants
11. Waste gum
12. Scrap metal
13. Waste glass and ceramics
14. Slag
15. Construction demolish waste
16. Excreta of animals
17. Dead animals
18. Dust

Present measures against industrial wastes

The treatment of industrial wastes are the responsibility of the persons who discharge these wastes. Under Article 3 and Article 4 of the Waste Disposal Law, industrial waste measures are taken respectively by enterprises, cities, towns and villages, prefectures and the State.

First, entrepreneurs tend to promote in-house treatment under the guidance of cities, towns and villages, as well as prefectures. That is, individual entrepreneurs are required to dispose of their wastes themselves. In some cases, they utilize an antipollution measures council in each prefecture as a place to exchange information on industrial wastes. Moreover, there are some moves toward establishing joint treatment operations through such consultative organs.

Meanwhile, some industries are promoting improved industrial waste measures. They are considering organizing an antipollution measures committee, working out antipollution policies, and establishing a nationwide treatment program. Measures under consideration include the treatment of slag in the iron and steel industry, waste tires in the tire industry, and used electric home appliances in the household electric products industry.

As for municipalities' measures against industrial wastes, cities, towns and villages are disposing of industrial wastes as well as domestic wastes in accordance with the provisions of Article 10-2 of the law. The landfill sites now established by cities, towns, and villages, are 2628 in number and 44 km^2 in area. Of the sites, those accepting industrial wastes are about 1% in number and about 7% in area. In the meantime, the cities designated by the Cabinet order, which are concerned with the subject of industrial wastes, are exercising guidance and supervision over entrepreneurs and waste treatment contractors. Some of them are making their own plans concerning industrial wastes.

Prefectures are taking various measures against industrial waste on the basis of industrial waste management plans as defined in Article 11 of the law. These measures can be broadly classified into surveys and research to develop disposal technology and collect basis data, the establishment of concrete processing (and recycling) plans based on the industrial wastes treatment program, the improvement of processing systems for troublesome regions or troublesome wastes and the guidance and supervision over enterprisers and wastes processing contractors.

These actions are promoted in a coordinated manner. The prefectures that established their industrial wastes treatment programs several years ago have recently been surveying the actual conditions of discharge of industrial wastes in an effort to review the treatment programs in response to economic and other changes. In addition, some prefectures are making basic surveys in preparation for industrial waste treatment operations.

Guidance and supervision over entrepreneurs and waste treatment contractors are a major part of each prefecture's industrial waste administration. Guidance and supervision are primarily conducted by environmental sanitation inspectors. As of April 1, 1985, they numbered about 3000 — a small number compared with the number of entrepreneurs and treatment contractors. It is difficult, therefore, to conduct adequate guidance and supervision.

Accordingly, prefectures are making efforts to ensure sufficient activities in these fields. To improve the effects of their guidance and supervising activities, they are conducting guidance and technical seminars for associations of entrepreneurs and treatment contractors. Meanwhile, exchange of information is becoming more important as the areas of industrial waste processing widen along with the scope of activities of treatment contractors. By organizing consultative groups prefectures are thereby exchanging information and studying common problems.

Under Article 4-3 of the law, the Ministry of Health and Welfare promotes the development of waste processing technology and provides technical and financial assistance for prefecture measures to grasp the conditions of industrial waste and to ensure proper processing of industrial waste. The ministry takes appropriate measures to achieve these ends. When local public bodies construct or improve facilities (including final treatment facilities) to dispose of industrial waste from sewerage treatment facilities and water purification facilities, the Ministry of Health and Welfare grants government subsides (subsidy rates: 25% for general region and 50% for antipollution plan regions). Also, in the metropolitan and Kinki regions, where the advanced use of land in recent years is making it particularly difficult to secure sites for landfilling, the ministry is promoting plans to construct final treatment plants, so-called "Phoenix plans," with involvement of the government.

History of hazardous waste management

The following legislation pertaining at least in part to hazardous waste management has been enacted:

Aug. 1967: Basic Law for environmental pollution control promulgated.
Feb. 1969: Second 5-year plan for Expanding Public Cleansing Facilities adopted at a Cabinet meeting.
Dec. 1970. Law governing waste treatment and public cleansing promulgated.
Apr. 1972: Pollution by PCBs aroused attention.
Jan. 1975: Third plan for Expanding Waste Treatment Facilities adopted at a Cabinet meeting.
Jul. 1975: Pollution by hexavalent chromium in Tokyo attracted attention.

Aug. 1976: Fourth plan for Expanding Waste Treatment Facilities adopted at a Cabinet meeting.
Apr. 1980: Office at Regional Planning set up in the Ministry of Health and Welfare.
Apr. 1981: Office of Industrial Waste Management set up in the Ministry of Health and Welfare.
Nov. 1981. Fifth plan for Expanding Waste Treatment Facilities adopted at a Cabinet meeting.
Nov. 1983. Dioxins and Furans detected in fly ash at a municipal waste incineration plant. Emission of mercury in dry batteries from municipal waste incinerators arouse attention. Separate collection of dry batteries is practised in some municipalities.
May 1984: Ministry of Health and Welfare gave an administrative guidance with respect to these problems.

Licenses/permits required

Article 15 of the Waste Disposal and Public Cleansing Law says:

> The one who is going to construct an industrial wastes disposal plant or to change its structure or scale (excluding small changes specified in the Ordinance of Ministry of Health and Welfare) shall give a notice to the prefectural governor in accordance with the Ordinance of Ministry of Health and Welfare.
>
> When the prefectural governor recognizes that the industrial wastes disposal plant in the notice in accordance with the stipulation in the foregoing Paragraph does not correspond to the technical standards specified in the Ordinance of Ministry of Health and Welfare (Ordinance of the Prime Minister's Office and the Ordinance of Ministry of Health and Welfare concerning the final disposal site of industrial wastes), he is entitled to order to change or to abolish the plan in the notice to the person who gave the notice only within 30 days from the date of the acceptance of the notice (60 days in terms of the final disposal site of industrial wastes).

There is no particular manifest system such as that undertaken in USA and West Germany; however, there is a kind of manifest system that makes three parties contract to ensure triangular cooperation between the generator, disposer and transporter in Japan.

Article 14 of the Waste Disposal and Public Cleansing Law says:

> Any person who intends to undertake collection, transport or disposal of industrial wastes as business, shall ask a permission of the business from the prefectural governor of the area in which he conducts the business. However, it is excluded, when the entrepreneur transports and disposes of the industrial wastes by himself, when a person undertakes collection, transport and disposal of only those industrial wastes which can be re-used or in other cases specified in the Ordinance of Ministry of Health and Welfare.

The prefectural governor shall not give the permission in the foregoing Paragraph, unless the application for the permission is in accordance with the technical standards specified in the Ordinance of Ministry of Health and Welfare.

The government of Japan declared on July, 1985, after careful observation and evaluation, that there is no specific environmental problem due to the incineration and/or disposal of waste batteries in the waste stream. However, the government recommended strongly to all those concerned that the following practices be considered:

- increased recovery rate of waste batteries by municipalities;
- alternative batteries be developed to replace existing mercury batteries;
- alternative batteries be developed specifically for electric equipment such as radios and calculators;
- alkali-batteries be coloured or otherwise marked by manufacturers;
- mercury in alkali-batteries be reduced by manufacturers.

The government also has recommended the following additional measures to all municipalities:

- provide treatment (disposal) system of waste alkali-batteries in each municipalities;
- provide inter-municipal network of recovery of waste alkali-batteries by municipalities, which will be supported financially by manufacturers;
- strengthen monitor systems for mercury uptake in exhaust gas and waste water from municipal facilities such as incinerators and disposal sites.

A Technology Development Center (TEDEC) was established as a sub-organization of the Japan Waste Management Association (JWMA) on November 22, 1985, under the full support by Battery Manufacturer Association for the purpose of helping municipalities with battery collection. TEDEC has arranged for a transportation company and a treatment company respectively to haul waste batteries stored in municipalities to the treatment company in northern Japan. TEDEC will keep track of disposal information, and report back to the municipalities.

Roles of governments and industry

The chief function of the national government is to promote research and development of waste treatment technology and give technical and financial assistance to regional and local governments.

The prefectural governors are responsible for preparing a distribution plan for industrial waste treatment plants, an area transportation plan, a plan for properly siting facilities, for keeping track of industrial wastes, and for taking the necessary action for proper treatment.

The prefectural governors receive reports from entrepreneurial industrial waste treatment contractors, and the managers of industrial waste treatment plants. They can inspect facilities, if necessary.

The prefectural governors issue permits for industrial waste treatment facilities, if the application is granted. The city and town mayors and village chiefs receive the prescribed reports from entrepreneurs.

No subsidy for hazardous waste treatment is granted by the national government, but the Public Nuisance Prevention Corporation and other groups offer loans for expansion of treatment facilities.

Industries may transport wastes for themselves or secure other qualified industrial waste treatment contractors with the prefectural governor's permit. In April, 1980, a total of 25,712 contractors had received permits to collect and transport industrial wastes.

Industries are required to keep their industrial waste in proper storage conforming to the storage guidelines until they are collected, transported and disposed of.

If they are to entrust collection, transport and disposal of industrial waste to another party, the producers should do so conforming to the guidelines. If they are to collect, transport and dispose of industrial waste with the permit, the transporters should do so, conforming to the given guidelines. They may entrust collection, transport and disposal of industrial wastes to another party, conforming to the entrustment guidelines.

Anyone who intends to construct an industrial waste treatment facility must notify the prefectural governor of that intention. The manager of treatment plans must operate their plants, conforming to maintenance and operational guidelines.

Use and manufacture of PCBs was prohibited in 1971. Currently manufacturers are required to remove parts containing PCBs from their goods, on their own responsibility. PCBs, if in prolonged storage, may pose a problem. Hence incineration of PCBs will be considered.

If small and medium-sized businesses and other parties, cannot treat and dispose of their wastes for themselves, the local governments set up the so-called "third sector" or a public corporation to meet their needs. Industrial wastes used in the past as land reclamation material pose environmental pollution problems, and are required to be cleaned up. Public clean-up action was taken recently in a case of soil pollution by hexavalent chromium in Tokyo.

HAZARDOUS WASTE DEFINITIONS

The purpose of defining hazardous waste in Japan is to support the prevention of water, land and marine pollution. The definitions require that hazardous wastes disposal on land be carried out at a place isolated from public waters and underground water. Such wastes should be treated before land disposal to a quality level conforming to the hazardous material extraction criteria, or solidified to prevent leaching of hazardous substances.

As for the hazardous wastes named in the first section of this report there is a special regulation in respect to mercury, cadmium, lead, hexavalent chromium, arsenic, cyanide, PCBs and other chloro-organic compounds. Explosive and radioactive substances, however, are not in the regulations. Also, these regulations are applied only to limited types of industrial waste containing these substances and their specified generators.

Industrial waste is generally conceived to be a single industrial waste matter discharged as a result of routine operations, but an industrial waste discharged in a mixed state may be regarded as a complex waste. An example of such waste is sulfuric acid pitch, which is a mixture of waste and waste oil.

SOURCES, QUANTITIES AND KINDS OF HAZARDOUS WASTES

There are no available statistics of hazardous waste generation in Japan. However, as far as industrial waste is concerned, the total generation of industrial waste has increased by roughly 23% from 1975 to 1980 as illustrated in Table 1.

TABLE 1
Generation Amounts of Industrial Wastes (comparison between 1975 and 1980, thousand tonnes per year)

Type of Wastes	Generation in 1980		Generation in 1975	
Cinder	1797	0.6%	1203	0.5%
Sludge	88,190	30.2%	37,660	15.9%
Waste Oil	2419	0.8%	2289	1.0%
Waste Acid	10,219	3.5%	9872	4.2%
Waste Alkali	6090	2.1%	14,435	6.1%
Waste Plastic	2232	0.8%	1480	0.6%
Waste Paper	1624	0.6%	991	0.4%
Wood Chips	6628	2.3%	7890	3.3%
Waste Rags	101	0.0%	204	0.1%
Animal Remnants	4323	1.5%	2596	1.1%

TABLE 1 (continued)

Type of Wastes	Generation in 1980		Generation in 1975	
Rubber Scraps	92	0.0%	597	0.3%
Metal Scraps	13,111	4.5%	9985	4.2%
Glass Chips and Porcelain Chips	2297	0.8%	2870	1.2%
Slag	60,561	20.7%	60,950	25.8%
Demolition Wastes	30,007	10.3%	34,144	14.4%
Livestock Excretion	49,629	17.0%	41,184	17.4%
Livestock Carcasses	62	0.0%	38	0.0%
Dust	11,731	4.0%	8101	3.4%
Others	1199	0.4%	—	—
Total	292,312	100.0%	236,489	100.0%

Some industrial wastes are defined as hazardous to the environment and human health if they exceed the prescribed hazard limits after undergoing an extraction test. Such hazardous wastes are restricted to ocean disposal unless treated by calcination for sludges containing mercury, and decomposition for cyanide-containing wastes. The following process plants treat such wastes:

Hazardous Waste Treatment Plants

Kind of plant	Entrepreneur	Contractor	Public body	Total
Concrete solidifying plant	55	44	1	100
Calcination for sludge containing mercury	3	4	0	7
Cyanide decomposition plant	373	19	18	410

Prior to disposal, hazardous wastes must be isolated to protect the public water and groundwater. To meet such requirement, they may be treated by concrete solidifying plant so that no hazardous leachate can result.

COLLECTION AND TRANSPORTATION SYSTEMS

Entrepreneurs should meet the following requirements when collecting and transporting industrial wastes: littering and spillage should not occur, and the transporting vehicle, container, and conveyance pipeline, must be structured to let out neither waste nor odor.

Some industrial wastes will be handled by the licensed contractors for treatment, collection and disposal. The number of such contractors is tabulated in Table 2. Most of these businesses are small-scale thus a more stable situation is needed.

TABLE 2
Number of Licensed Contractors for Treatment, Collection and Disposal of Industrial Wastes (1980)

Collection and transportation only	23,226
Processing only	264
Disposal only	171
Collection, transportation and processing	1393
Collection, transportation and disposal	843
Processing and disposal	41
Collection, transportation, processing and disposal	250
Total	26,188

Standards for hazardous waste disposal and management

Public sewerage sludge, or such materials treated for disposal, and ash, dust or cinder that contains mercury and its compounds or cyanide compounds, or other materials judged to be toxic are controlled in the following ways:

- if these materials are solidified in concrete, landfilling is required so that there is interception of any seepage into groundwater and public water areas;

- if these materials are not solidified in concrete, they must be detoxified and disposal of in general landfills.

Public sewerage sludge, or these materials treated for disposal, and ash and dust or cinder that contain cadmium and cadmium compounds, lead and lead compounds, organic phosphate compounds, cr^{6+} compounds, arsenic and arsenic compounds, are therefore judged to be toxic. Also, slag or slag treated for disposal, which includes five kinds of toxic materials, are also judged to be toxic. They may be rendered non-toxic and landfilling may occur, if possible seepage from the landfill is prevented.

When sludge landfilling occurs, the following standards are prescribed, regardless of whether materials are toxic or non-toxic:

- in cases where sludge landfill disposal occurs (excluding wet-type landfill), prior incineration using incineration equipment must occur, or the water content must be below 85%;

- in cases where organic sludge (sludge removed from public sewers or river basin sewers, excluding materials digested using digestion facilities and those materials whose organic content is smaller than that of materials digested using digestion facilities) is disposed of by wet-type landfill, prior incineration using incineration equipment must take place;
- for organic sludge or such sludge treated for disposal containing more than 40% organic wastes (excluding materials incinerated with an ignition loss of less than 15%, and those solidified in concrete) each landfill layer should be less than 50cm thick. In cases where organics are less than 40%, each layer should generally be less than 3m thick.

Moreoever, each layer should have a surface cover of about 50cm of sand and soil. In cases of small-scale landfill disposal (landfill in an area of less than 10,000m^2 or of a volume less than 50,000m^3), however, or in cases of landfill disposal utilizing subterranean air spaces, this is not to be applied.

Also, along with the provision of ventilation equipment for the landfill area and the removal of methane and other gases produced in the area, measures necessary to prevent the outbreak of fires should be practised. But in cases of small-scale landfill disposal this is not to be applied.

When landfilling of industrial wastes occurs, the following standards are applied:

- Necessary precautions should be taken to prevent offensive odors from emanating from the landfill site.
- Rats should be prevented from living on landfill sites, and mosquitoes, flies, and other harmful insects should be prevented from reproducing.
- Landfill sites should be enclosed with a fence, and the fact that they are areas for the disposal of industrial wastes should be indicated. Hazardous industrial waste landfill sites also should be marked.
- Landfill sites should be insulated to protect public river basins and groundwater. But in cases of disposal of non-hazardous industrial wastes, when the necessary precautions are taken to prevent contamination of public river basins and groundwater by landfill leachate, this is not to be applied.
- Landfill disposal of acid and alkaline wastes is prohibited, regardless of whether they are hazardous or non-hazardous.

Standards for ocean dumping apply in the following ways:

- Sludge or authorized public sewerage sludge, slag, or acid or alkaline wastes, ash and dust or cinder judged to be hazardous must not be disposed of in the oceans.

8. JAPAN

- Sludge or authorized public sewerage sludge, which includes cyanide compounds is judged to be hazardous, but if it has been incinerated with less than 15% ignition loss it can be disposed of in Marine Area B.
- When sludge or authorized public sewerage sludge judged to be hazardous is solidified in concrete so that toxic materials included in the sludge will not leak, it may be disposed of in Marine Area A.
- When sludge or authorized public sewerage sludge includes mercury and mercuric compounds and is judged to be hazardous, but has been calcinated and thereby made non-hazardous, it may be disposed of in Marine Area B.

Detoxified inorganic sludge (excluding water-soluble materials) and detoxified slag also may be disposed of in Marine Area B. Detoxified organic sludge, detoxified water soluble inorganic sludge, or detoxified acid or alkaline wastes may be disposed of in Marine Area C.

Some sludges and acid or alkaline wastes may not be disposed of in the oceans if the materials are types of oils, or if the materials are discharges from a phenolic resin manufacturing industry and contain phenols.

The standards concerning methods of discharge when discharging wastes in any of the marine areas are given in the following chart:

Discharge area	Standards for discharge method	Related standards	
Marine Area A	1. Specific gravity must be over 1.2 when discharging. 2. Discharge must not take place when vessel is moving.	Precautions must be taken which are necessary to ensure the wastes sink as quickly as possible, and that they accumulate as sediment.	Endeavor to avoid places where there is a fear of hindering the growth of marine flora and fauna.
Marine Area B	1. and 2. from above, and: 2. No discharge in powder form.		
Marine Area C	1. Discharge below the surface of the ocean. 2. Discharge while the vessel is moving.	Discharge waste in small quantities at a time Take necessary precautions to ensure that the wastes diffuse in the ocean as quickly as possible.	

In general, even in the cases of industrial wastes that can be disposed of in the oceans, ocean dumping should not occur if landfill disposal is available.

Collection, transportion, and disposal of wastes should be carried out so that the wastes will not fly about or drift away. Likewise, waste treatment facilities should be established to conform to the rules of environmental preservation.

Finally, precautions should be taken so that there is no fear of wastes flying about, drifting away, or giving off offensive odors from transportion vehicles, receptacles, or pipelines.

As to the collection systems, waste oil is not hazardous waste in Japan, but is transported to oil recovery company by contractors. Pipeline collection for domestic and partially commercial wastes are in operation in a few cities. Generally, municipal wastes are the responsibility of local public bodies, and industrial wastes are the responsibility of entrepreneurs. Neither of these is systematically handled, however. Industrial waste treatment contractors have formed a nationwide organization, and have been making efforts to increase the technical level and promote proper treatment of industrial waste.

TREATMENT, STORAGE AND DISPOSAL SYSTEMS

Table 3 shows the hazardous waste treatment facilities which exist in Japan.

TABLE 3
Industrial Waste Treatment Plants (number of facilities)

Kind of plant	May,'77	May,'78	Apr.,'79	Apr.,'80
Sludge dehydration plant	2,125	2,760	3,096	3,446
Sludge drying plant	174	178	187	203
Sludge incineration plant	356	413	409	427
Oil-water separation plant for waste oil	535	499	322	447
Incineration plant for waste oil	312	361	375	397
Waste acid or alkali neutralization plant	684	623	356	452
Waste plastics crushing plant	77	87	95	105
Waste plastics incineration plant	641	786	879	1,064
Concrete solidification plant	86	95	93	100
Calcination plant for mercury containing sludge	4	5	9	7
Cyanide decomposition plant	349	316	317	410
Strictly controlled landfill sites	—	25	25	21
Least controlled landfill sites	—	170	207	217
Controlled landfill sites	—	599	756	532
Total	—	6,917	7,126	7,828

Hazardous wastes are treated or disposed of only at concrete solidification plant, calcination plant for mercury containing sludge, cyanide decomposition plant, or strictly controlled landfill. Wastes, basically, are handled as shown in the flow chart.

Flow Chart of Industrial Waste Treatment (1980)

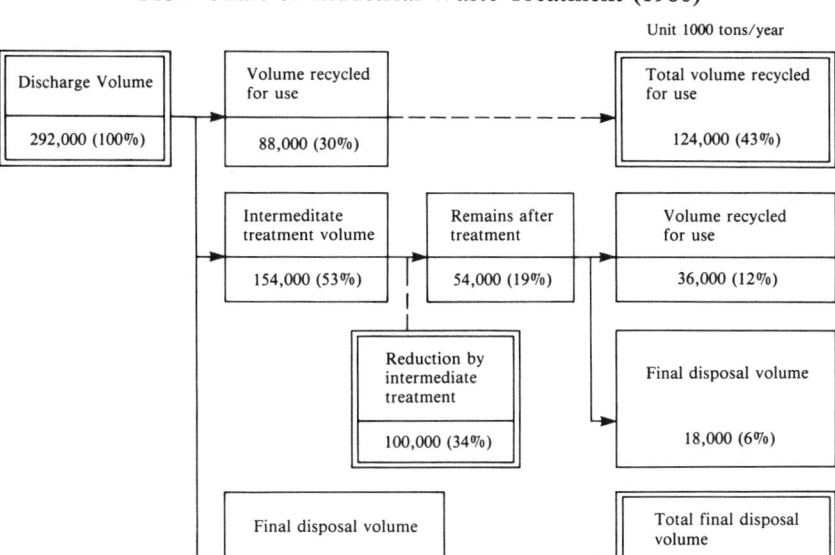

Approximately 43% of the total quantity of industrial wastes generated throughout Japan in a year is recycled. Examples of recycling are given in Table 4.

TABLE 4
Examples of Recycled Industrial Waste

Kind of waste	Instances of recycling
Organic sludge	Fertilizer
Inorganic sludge	Raw material for cement and aggregate
Waste oil	Regeneration
Waste acid	Generation of gypsum, ferrous sulfate, sulfuric acid band, etc. from waste sulfuric acid and recovery of chromic acid and bichromate acid from the waste water from a chrome-plating process
Waste alkali	Generation of calcium carbonate and aluminum hydroxide
Waste plastics	Regenerated plastics and energy recovery
Metal scrap	Recovery and reuse

TABLE 4 (continued)

Kind of waste	Instances of recycling
Glass chips	Recovery and reuse
Vegetable and animal remnants	Feed made from the waste water after food processing and generation of condiments
Waste paper	Regeneration
Wood chips	Regeneration
Waste rags	Production of non-textile cloth with a synthetic fiber and asphalt sheet
Slag	Raw material for cement (steel slag) and roadbed material
Construction and demolition waste	Use of roadbed concrete and regeneration of asphalt
Livestock excretion	Fertilizer
Dust	Fly ash recycling and recovery and vanadium oxide

The volume of waste generated is reduced by renovating and improving industrial structures and processes. Since industrial waste is quite varied from industry to industry, development of recycling technology is required in addition to efforts on the part of individual businesses to reduce the amounts of industrial waste. Currently, the national campaign, "General Researches on Waste Treatment and Technology for Converting Wastes into Resource," is being undertaken by research and development interests in Japan.

Ownership of disposal facilities is indicated below:

TABLE 5
Industrial Waste Treatment Plants

Kind of plant	April 1, 1985			
	Entrepreneur	Contractor	Public body	Total
Sludge dehydration plant	3911	223	690	4824
Sludge drying plant	105	28	41	174
Sludge incineration plant	359	89	56	504
Oil water separation plant for waste oil	167	148	2	317
Incineration plant for waste oil	296	145	2	443
Waste acid or alkali neutralization plant	221	18	0	239
Waste plastics crushing plant	52	114	4	170
Waste plastics incineration plant	950	382	17	1349
Concrete solidifying plant	40	52	3	95
Calcination plant for mercury containing sludge	1	5	0	6

TABLE 5 (continued)

	April 1, 1985			
Kind of plant	Entrepreneur	Contractor	Public body	Total
Cyanide decomposition plant	256	33	22	311
Strictly controlled landfill sites	21	11	0	32
Least controlled landfill sites	103	578	19	700
Controlled landfill sites	342	406	107	855
Total	6849	2240	1001	10,090

Probably no hazardous wastes are transported to other countries for disposal, since Japan is an island country. As far as we know, there is no future possibility of transporting hazardous waste to other countries for disposal.

ASSESSMENT

An important regulatory tool for the management of hazardous waste was adopted in the form of the Japanese "Environment Impact Assessment" (EIA), by the Japanese Cabinet at its meeting on August 28, 1984. Almost all facilities are now planned with the assistance of an EIA.

The national government plays the key role in control system implementation in Japan. It gives guidance to entrepreneurs about the specific permits needed to carry out their operations and in so doing it seeks the cooperation of entrepreneurs and local public bodies. It also monitors the specific actions of local public bodies to match them with the national guidelines regarding EIA.

The following steps are required for control reviews:

- formulating a statement on the preparations for assessing impact on the environment;
- making this statement known;
- giving an opinion on this statement;
- preparing a statement of the assessment of the impact on the environment.

Expectations for the future include the enforcement of statutory regulations. The general regional developmental plans will include a waste treatment plant project. These plans will be executed on the basis of general public need.

The number of environment and sanitation professionals will be increased to ensure close supervision, enlightenment and further spread of knowledge. Public body participation and commitment will be enhanced.

The need for research and development will increase, and further education and training of individuals will take place.

Areas in need of special attention are the proper control of wastes from medical institutions; siting of landfills; the promotion of area waste-treatment plants; and the upgrading of information control systems.

9

Hazardous Waste Management in THE NETHERLANDS

D. den OUDEN

AVR Chemie NV, Rosenburg, The Netherlands

OVERVIEW

In The Netherlands there is a broad range of industrial activities and in this high populated country toxic waste is produced in many places. At the moment there is insufficient information available concerning chemical waste. There is a great need for collection of data to compile an inventory of the mode of production, frequency of production, amount produced, type and origin of the waste.

There is a framework of acts on the environment in The Netherlands. The requirements for licensing, the limited periods for which permits are granted and the conditions attached to permits encourage ways of preventing waste. In the last 20 years many industries with high impact on the environment have been reorganized. Use of materials which can produce toxic waste, like PCB or asbestos, is forbidden or allowed only under stringent conditions.

Hazardous waste has been landfilled in the past, although a rotary kiln incinerator was started in 1973. A nationwide research program has identified many places which are potentially contaminated. Recuperation of the contamination takes years. New industrial activity has been started to clean all these sites. Techniques like biodegradation, detoxification by washing and incineration are used. The clean-up of Lekkerkerk is an example known worldwide.

There are various recuperation and disposal facilities which are normally small plants. These private enterprises serve part of the market. For toxic waste without an economic value, disposal is centralized. A company, AVR-Chemie CV, has to operate competitively within the Dutch marketplace. The share holders of this company are the national government (10%) the city of Rotterdam (45%) and eight multi-nationals (45%).

Main activities at this facility are high temperature incineration, controlled landfill and management of small quantities.

The first priority for the national program is waste prevention. This entails first preventing the production of chemical waste and secondly promoting the reuse and recycling of such waste in preference to disposing of it as such. In addition, prevention in the broader sense involves cambating environmental pollution caused by the final disposal of waste, notably by incineration and disposal at sea and on land.

A second main priority is to combat the discharge of chemical waste and waste oil into the environment. For this purpose, efforts are being made to indicate ways in which these substances can be disposed of effectively in an environmentally acceptable manner. Every effort is being made in relation to both priorities to control both the quantity and the quality of waste and to gain more information about the problems caused by chemical waste and waste oil.

Priority must be given to preventing the production of waste. However, it must inevitably be accepted that some waste is bound to be produced. As new processes and products are introduced, new types of waste take the place of old. In fact even environmental protection measures intended to reduce emissions of pollutants into water, air and soil produce new wastes. Thus, controlling waste is both a short-term and long-term aim. It is most important to see policy on waste in conjunction with policy on protecting the said features of the environment.

NATIONAL CONTROL SYSTEM

Summary of legislation

In The Netherlands the synonym for toxic waste is chemical waste. The Chemical Waste Act (Wet Chemische Afvalstoffen) was issued on June 13 1979. An indicative multi-year program for chemical waste has been published for the years 1985–1989.

The main policy intentions for the years ahead are:

- to obtain more information about chemical waste,
- fuller measures to prevent chemical waste,
- the establishment of basic facilities for disposal of chemical waste by

means of incineration and land disposal and of channels leading to disposal facilities, particularly for smaller quantities of waste, mostly generated by smaller business,
- efficient division of responsibilities and powers among all levels of government, involved in implementing chemical waste policy and improving chemical waste management,
- improving the capabilities of these authorities.

The Chemical Waste Act itself requires the following improvements during the years ahead:

- the Substances and Processes Decree persuant to the Chemical Waste Act will be amended in order to simplify the application of the Decree in practice and to make it easier to recognize chemical waste,
- the act will be amended to permit the Minister to issue instructions to individual businesses. A provision will also be incorporated in the act so that the mixing or dilution of chemical waste with other substances or wastes can be limited in specific cases,
- the current notification system will be altered; in future notifications concerning transfers of chemical waste by the producer must be sent to the provincial authorities,
- a 4-year program is being drawn up in order to improve enforcement,
- the EEC directive on the transboundary shipment of dangerous and toxic waste will be implemented.

In the last 10 years many revelations of malpractice have provoked public opinion. Several firms have been closed because of these, and many places are still contaminated. Because of the mix of private and semi governmental enterprise involved, the whole waste treatment field was under suspicion. In the late 1970s, together with the implementation of the Chemical Waste Act, a system of licenses and permits was introduced for facilities handling or reworking chemical waste.

Licenses and manifest systems

At the moment every chemical waste treater is obliged to have a permit for his activities and is well controlled by the authorities. The Chemical Waste Act is a national act. Permits for a certain activity are only given when the activity is in line with the national waste management view of central government.

The producer is responsible for the waste until the acceptance of the waste by a permit holder; the transporter has no responsibilities. The EEC directive on the transboundary shipment of dangerous and toxic waste will be implemented in the near future.

Research for ways of reusing or recycling waste can be subsidised. The efforts to establish new investments for expansion of the basic facilities for final disposal (incineration in rotary kiln and controlled landfill in a concrete basin) have been subsidised.

Transport systems

In The Netherlands chemical waste is mostly transported by road carriage. The transporter has no special responsibilities and the transport of waste is similar to the transport of original products.

Responsibilities

The generator is responsible for the waste until the acceptance of the waste by a permit holder. Every transfer has to be notified to the central government. The transporter has no responsibilities. The treatment disposal facility is allowed by permit to execute certain activities and to accept chemical wastes with a composition within the regulations of the permit. Every transfer must be notified to the central government.

The Chemical Waste Act is a national act. The environmental control of the activities is the responsibility of provincial or municipal authorities.

Programs for dealing with the old, closed, or abandoned sites

To clean-up abandoned sites is an expensive problem. The principle that the polluter pays is often not applicable because the polluting firm is no longer in existence. In this case national government pays 90% of the clean-up costs and the local authorities the remaining 10%. A nationwide inventory program of contaminated places is almost finished and a clean-up program is being developed.

DEFINITION, SOURCES AND QUANTITIES OF HAZARDOUS WASTE

In The Netherlands there are two acts for regulation of waste. The Normal Waste Act (Afvalstoffenwet) regulates non-hazardous waste, like domestic waste, rubbish etc. The Chemical Waste Act (Wet Chemische Afvalstoffen) regulates hazardous waste. A precise definition of chemical waste is almost impossible. In the Chemical Waste Act a two-way description has been chosen:

- all toxic components have been listed in four categories of concentration levels (50 mg/kg, 5000 mg/kg, 20,000 mg/kg and

50,000 mg/kg): above these levels a substance is automatically chemical waste,
- a list of processes have been established and when a waste is generated from these processes then it is automatically a chemical waste.

Every chemical waste transfer must be notified to the ministry. Thus, on a national level a good overview is possible of the total amount of chemical waste produced and of its final destination.

The Substances and Processes Decree covers most hazardous waste, and at the moment the decree is under evaluation for amendment. Substances will remain covered or will be subject to the decree where this is in the interests of the environment. In particular, attention is paid to the highly toxic substances with concentration levels under 50 mg/kg.

All waste which is subject to the Chemical Waste Act is treated equally and the same notification procedure has to be followed.

It is estimated for 1985 that around 1 million tonnes of chemical waste and over 80,000 m^3 of waste oil was produced in The Netherlands. The disposal of 500,000 tonnes of chemical waste and 10,000 m^3 of waste oil took place in the producers' own facilities (inside the gate). The disposal of some 500,000 tonnes of chemical waste and 70,000 m^3 of waste oil fell under the provisions of the Chemical Waste Act (outside the gate). Since 1984 an enormous program has been under execution by the central government to quantify the remaining unknown amount of chemical waste (1984 estimation: 100,000 tonnes).

Of the 500,000 tonnes of chemical waste covered by the act, in total, 418,000 tonnes of chemical waste were registered and approximately 100,000 tonnes were estimated to be unknown in 1984. Also in 1985, the transfer of 39,000 tonnes of chemical waste and 44,000 tonnes of waste oil was reported. Table 1 presents the methods of disposal for the 418,000 tonnes of chemical waste registered in The Netherlands in 1984.

TABLE 1
Methods of Waste Disposal in The Netherlands (1983–1984)

Method	*Amount* (tonnes)
Distillation	12,400
Detoxification	25,600
Dewatering	175,700
Incineration on land	108,600
Incineration at sea	2200
Miscellaneous	15,500
Controlled landfill	71,000
Sea dumping	7000
Total	418,100

COLLECTION AND TRANSPORTATION SYSTEMS

The transporter has no special responsibilities; the same regulations are valid as applicable for the original products. The generator and the permit holder must notify every transfer of chemical waste to the central government. No special license system is in force for transporters.

Because of the implementation of the EEC directive on the transboundary shipment of dangerous and toxic waste a trip ticket system will be introduced.

The transportation and collection of toxic waste is normally in the hands of private enterprise. Special equipment for vacuum cleaning etc. is often needed in combination with maintenance assistance to collect the waste.

Waste oil collection system

The collection and permit system pursuant to the Chemical Waste Act operates satisfactorily. The permit holder is obliged to collect waste oil in his area, when a generator has more than 400 litres per year. Approximately 80,000 m^3 of waste oil are collected.

Recently some revelations about PCB contamination have shown that monitoring of the quality of waste oil is required, as it is not uncommon for liquid chemical waste to be mixed with waste oil. Quality requirements for processing will therefore be tightened up.

Household hazardous wastes collection systems

In 1987 a project will be organized to make efficient arrangements for the collection of the small quantities of chemical waste produced by numerous widely dispensed sources throughout The Netherlands. Due account will be taken of existing initiatives in the field of collection.

Regulations will be drawn up for collection of chemical waste from shipping. Every 1 or 2 years a national collection program is executed for the collection of insecticide residues and about 150 tonnes of insecticides are collected.

The collection system for household hazardous waste is in the midst of implementation. At the moment many different systems have been chosen by municipalities or provinces, but they have not been in effect long enough to enable them to be evaluated.

TREATMENT, STORAGE AND DISPOSAL SYSTEMS

Storage systems

Brokers are active on the Dutch market. Normally the broker has small facilities for storage of chemical waste. The brokers are intermediaries between the generator and permitted facility, or intermediaries between the generator and a treatment facility abroad. The storage of chemical waste is done on the site of the producer or on the site of the treatment facility.

Incineration systems

Incineration of chemical waste is done in several installations in The Netherlands. The large sites of the chemical industries often have their own facilities which are normally not open for other customers.

In 1973 the incineration of chemical waste in rotary kilns was started. In 1984, 45,000 tonnes of chemical waste were incinerated in a rotary kiln installation. A new rotary kiln incinerator is under construction with a capacity of 40,000 tonnes and came on stream in the second half of 1986.

A great part of domestic waste is incinerated in The Netherlands. In some of these installations specially prepared chemical waste is co-incinerated.

Incineration at sea has decreased in the last few years and it is expected that it will no longer be allowed when the new rotary kiln is fully operational.

In total about 110,000 tonnes of the generated chemical waste are incinerated per year, of which 40,000 tonnes are incinerated in rotary kilns, 40,000 tonnes are co-incinerated with domestic waste, 2000 tonnes are incinerated at sea and the remainder is exported.

Landfills and disposal of waste on the land

Disposal of chemical waste on land, underground or at sea is forbidden although in special cases exceptions are made.

Some landfills for domestic waste and rubbish are still in operation, although many are closed. On these landfills it is forbidden to dispose of chemical waste. For some toxic waste exceptions have been made to co-landfill under special regulations and arrangements.

Efforts to establish a controlled landfill for non-processable chemical wastes seem to be successful. Agreement has been reached for the design and location of the landfill. Depending on the permit procedure it is expected that it will be operational in 1988 or 1989. At the moment 80% of the chemical waste for landfill is exported.

In The Netherlands there are no existing facilities for underground disposal in mines or caves. In the east of The Netherlands there are underground salt mines and possible usage is now being researched.

Underground injections into deep wells is forbidden. No permit holders are known for surface impoundments such as pits, ponds and lagoons.

Co-disposal with refuse is normally forbidden but exceptions are given in a few cases. In total 12,000 tonnes of chemical waste are landfilled in The Netherlands, most of this landfill is executed as co-disposal.

Sludge from biological treatment plants has been used for many years as a fertilizer. In the last few years regulations have been issued for the content of heavy metals and chlorinated organic compounds. The sludge from an industrial biological treatment plant is by definition chemical waste. Most of these sludges are no longer allowed to be used as fertilizers. After dewatering the sludge is incinerated or landfilled.

In 1984 about 7000 tonnes of waste were disposed of at sea. There is a great opposition to sea disposal. Permits are given for short periods under the conditions that better solutions have to be found within a certain period.

Energy recovery and recycling

Almost half of the production of toxic waste (500,000 tonnes per year) is reworked by the generators at their own faciltiies. A great part of these streams is used as feedstock for production, as fuel for energy recovery or is redistilled. The remaining part (estimated 50%) is disposed of — for example by incineration without energy recovery or bio-degradation. About 10,000 tonnes is distilled by special waste treatment firms. The products are sold on the free-market.

In The Netherlands there is a nationwide collection system by private enterprises for silver recovery from photographic material.

Description of programs to reduce waste generation

Research into clean or environmentally acceptable technology involving modified processes and modes of production and alternative raw materials will be continued. Research findings will be communicated to companies individually, via organizations representing the various sectors of industry, via an information center, which is to be set up by industry, central government, and chambers of commerce and industry.

Within the framework of licensing under the Chemical Waste Act and the Marine Pollution Act, the limited periods for which permits are granted and the conditions attached to some permits will encourage research into ways of preventing waste.

Further research will be conducted into how the production of chemical waste can be further prevented by means of a coherent policy on producing hazardous substances.

Operation and ownership of facilities

About 40 permit holders are allowed to handle chemical waste. These permit holders are mostly small firms who specialize in a certain part of the market. They are privately owned with a few exceptions. Most of these firms are organized into branches of the national chemical association.

The basic facilities for final disposal of chemical waste are centralized. The AVR-Chemie is jointly owned by central government (10%), city of Rotterdam (45%) and eight multi-nationals (45%).

The main activities at this facility are:

- two rotary kilns (capacity 90,000 tonnes per year),
- co-incineration on moving grid furnaces (capacity 70,000 tonnes per year),
- controlled landfill (capacity 230,000 tonnes, construction),
- special activities like collection of small quantities, storage of contaminated soil.

Because of the open market in The Netherlands the AVR-Chemie has to be competitive with other existing disposal facilities in The Netherlands and abroad. The average charge is DFL 300 per metric tonne of toxic waste.

Imports and exports

In 1984 approximately 120,000 tonnes of waste were exported: Belgium (20,000 tonnes), FRG (38,000 tonnes), UK (5,000 tonnes), France (20,000 tonnes) and GDR (32,000 tonnes). At present one rotary kiln is under construction by AVR-Chemie and also a controlled landfill. When these facilities come on stream The Netherlands will be, in principle, self-supporting. Export of chemical waste will then be done for specific treatment or for economic reasons only.

ASSESSMENT AND FUTURE DIRECTIONS

A 4-year program is being drawn up in order to improve enforcement. This project is intended to bring about a considerable improvement in the observance of legislation and regulations on chemical waste and to create a situation in which provincial, municipal and other authorities will be in a position to take upon themselves much of the responsibility for implementation.

Every transfer of chemical waste has to be reported by the generator and the treatment firm. In 1984 it was estimated that 100,000 tonnes of chemical waste were not reported. The improvement of enforcement must eliminate this unreported waste. The notification system has been computerized and overall information will now become available about qualities and quantities.

Recent revelations of malpractice have shown that chemical waste is still exported under a fancy product name although the EEC directive on transboundary shipment gives more possibilities for control.

Dilution of waste hydrocarbons in waste oil is difficult to control. Only improvement of enforcement can eliminate these practices. Small quantities of hazardous waste are still co-disposed with other hazardous materials.

The Substance and Processes Decree pursuant to the Chemical Waste Act is difficult for non-professionals to understand. The decree will be simplified to make it easier to recognize chemical waste.

The future collection and classification of data will provide material of great value in defining possibilities for recycling, reusing and the efficient disposal of chemical waste. In the future there will also be a more efficient division of responsibilities and powers among all levels of government involved in implementing chemical waste policy. The chemical waste management capabilities of these authorities will also be more controlled.

The Chemical Waste Act has been in existence since 1979. At that time the basic facilities for final disposal were inadequate. Many companies could not dispose of their waste within the regulations of the act. In the beginning of 1980 much malpractice become known.

Since the implementation of the act, the disposal of chemical waste has become easier to trace, particularly by central government. However, the Chemical Waste Act has been implemented under the responsibility of central government without efficient division of responsibility between provincial and municipal authorities, and enforcement has not been as strict as it could have been. Some corrupt companies have taken great advantage of this.

Special attention is now paid to the small quantities of chemical waste generated throughout our society. A major problem is how to collect and treat waste at an acceptable price. Waste oil can be contaminated with highly toxic material, like PCBs. Recent revelations show the need to enforce regulations. The Marpol agreements, concerning the contamination of the sea by residues from tankers, have to be implemented in the second half of the 1980s. Many contaminated abandoned sites must be cleaned up. The implementation of regulations concerning highly toxic components at low concentrations (less than 50 ppm) needs more attention.

10
Hazardous Waste Management in SOUTHERN AFRICA

A. C. de BRUIN*

Member, City Council, Cape Town, South Africa

OVERVIEW

Introduction

In the context of this report, Southern Africa is defined as the subcontinent south of the Kunene and Zambesi Rivers. The region comprises at least 17 sovereign states and several geopolitical areas. The region is characterized by a mixture of first and third world levels of development. The two factors common to the entire region are a colonial background and a rapid change in political and other structures. In fact, the current situation highlights the profound wisdom of Prime Minister Harold MacMillan's famous "winds of change" speech at the Cape in February, 1960. Superimposed on incredible demographic changes are major social and political developments, which will determine the future development of the region.

The region has a fast growing population and a relatively low per capita income. The population of 68 million is approximately 25% of that of the EEC countries, but the Gross National Income of 85.4 billion US dollars is only 3.2% of that of the EEC countries. The value of mineral production in

---

(*) Mr de Bruin serves as Chair of the Southern Africa Working Group on Hazardous Waste. Other members include: J. J. Malan and J. R. Hale, Council for Scientific and Industrial Research, Pretoria; R. Lombard, Consultant, and S. J. Verrier, City Council, Johannesburg.

1973 as a percentage of the gross domestic product and mineral value per capita are indicated in Table 1.

The countries of Southern Africa differ widely in so far as area and population are concerned. Apart from Angola and South Africa, which each cover an area of more than 1 million km^2, there are relatively small states such as Venda (6113 km^2) and Swaziland (17,356 km^2). Larger areas such as SWA/Namibia and Botswana are practically uninhabited and have an average population density of only 1 person/km^2.

When looking at the map of Africa it is easy to overlook the fact that Southern Africa, were it as densely populated as the EEC countries, would be carrying a population of approximately 1 billion instead of the present 68 million.

With the exception of South Africa, and to a degree Zimbabwe and SWA/ Namibia, all the countries of this region display the characteristics of less-developed countries — first, a rapidly increasing population with low levels of productivity caused by illiteracy, a lack of training, poor health conditions, the retention of a pre-industrial value system as well as a shortage of capital and of management skills; second, extensive economic dependence on one or a few agricultural products or minerals; third, dependence on foreign financial and technical assistance; fourth, an increasing inability of agricultural and food production to keep up with population growth with the result that increasing quantities of food have to be imported but with growing malnutrition and famine; and fifth, a physically inferior infrastructure.

Because of the foregoing, industrial growth has taken place in countries like SWA/Namibia, Zimbabwe, Botswana, Swaziland, Lesotho and South Africa. South Africa and its immediate neighbors are the most developed in this respect. Therefore hazardous wastes are most prevalent in these states.

In South Africa there is a growing concern within the private and the public sectors about the build-up of harmful wastes. Because of this, corrective action has been taken with increasing momentum over the last 30 years. This includes improved legislation, stimulation of research and the accumulation of data on the latest technology and the development of original technology.

The challenge for the Institute of Waste Management (Southern Africa) is to obtain and develop appropriate technology and the transfer of this technology throughout the region.

In 1984, the institute arranged a seminar in Leboa with the aim of introducing the basic concepts of proper solid waste management. The seminar was attended by people directly involved in the cleansing branches of local authorities in Leboa and the surrounding areas.

The topics covered on this occasion were generation, collection, disposal, environmental and health aspects, costing and economics. The presented papers exposed the hard facts and basic concepts of solid waste

TABLE 1
Value of Mineral Production in 1973 as Percentage of Gross Domestic Product and Mineral Value *per capita* in Southern Africa

Country	Value		Value of mineral production R million	Mining sector as % of GDP	Value per capita (R)	Per capita	
	Rank in Africa	Rank in world				Rank in Africa	Rank in world
Botswana	21	96	22.1	31.6 (1980)	35.1	13	53
Lesotho	37	139	0.3	3.5 (1978)	0.2	35	134
SWA/Namibia	8	53	218.0	47.7 (1978)	255.8	2	13
South Africa (1)	2	10	3225.3	23.1 (1980)	129.6	5	24
Swaziland	20	94	24.3	4.0 (1977)	52.4	11	39
Southern African Customs Union			3490.0				
Angola	7	42	295.4	15.2 (1973)	52.8	10	38
Malawi	40	150	0.1	0.1 (1976)	0.1	44	150
Mozambique	33	119	4.5	0.2 (1973)	0.5	33	125
Zambia	5	25	816.4	17.0 (1980)	171.9	4	20
Zimbabwe	11	58	191.8	5.2 (1978)	31.4	15	59
Southern Africa			4798.2				

(1) Figures for Bophuthaswana, Ciskei, Transkei and Venda included under South Africa.

management to the delegates in a very efficient manner. Proof of this was the fact that each presentation was followed by a lively discussion.

A seminar was also organized in Swaziland. On this occasion the following topics were highlighted in well prepared talks: equipment, vehicles, legislation and control, siting and design of disposal sites, management of disposal sites, hazardous wastes and health risks, costing systems, and recycling.

The state of the art of waste management in Swaziland can be summarized by the following:

- only one town has a compaction vehicle;
- no specially designed machinery is being used because it does not seem practical since the second largest town processes only $25m^3$ per day;
- specially designed disposal sites are non-existent, currently dongas (erosion furrows) are being used and these are filled only to ground level;
- litter containers do not exist;
- no costs structures exist, waste handling expenses are being covered from the water tariffs;
- there is no legislation at all; and
- it seems that the cleansing departments do not even have the capacity to handle all the waste produced since large waste producers were refused services.

In Lesotho the main problem seems to be littering and harmful effluents produced by industry. Large quantities of food wastes also appear to produce certain problems.

Landfilling is currently being practised as far as it entails the filling up of dongas and worked quarries. There is, however, a general need for the correct equipment. The officials involved are aware of the dangers of groundwater pollution and are taking some measures to prevent it. A study in which surface waters were examined for the presence of cadmium, copper, chromium and lead, was recently done. The outcome of this was that the heavy metal pollution was negligible. Incineration is utilized mainly for hospital wastes.

Some legislation does exist, but it tends not to be specific enough. Currently there are some industrial developments responsible for the normal production of harmful effluents. Chemical products, as well as chemical wastes, are also transported within the country without any control by means of legislation. A serious problem exists due to the presence of hazardous wastes in domestic refuse bins. Because of this, chemical poisoning to children and domestic animals has occurred.

In Harare, Zimbabwe, the bulk of hazardous waste comes from the following sources: metal processing industries, tanneries, oil refineries, asbestos manufacturers, chicken farming, and other industries. These wastes are being disposed of in quarry area landfill sites where hazardous wastes are admixed and no records kept. No law enforcement to the effect of sound waste management practices is being applied although a fairly satisfactory system of controlled disposal exists within the city.

After years of uncontrolled disposal the position of a Noxious and Toxic Wastes Officer was created to control the disposal of hazardous wastes in Harare. A permit system was devised together with a scale of charges. The identification of hazardous wastes is an ongoing process.

The cost of refuse removal in Balawayo can be summarized as follows:

Type of waste	Cost per m^3 (US dollars)
Garden refuse	2
Light industrial	3
Heavy wastes	4

Regarding waste management, all the states in Southern Africa have much in common. The main problem areas are:

- littering of mainly beer cans and plastic packaging;
- no control whatsoever over the handling and disposal of hazardous wastes; and
- a lack of proper equipment to properly operate the existing landfill sites.

The Southern African Institute for Waste Management is currently very active in organizing training seminars in all states of Southern Africa. The response to this has been very positive and the feeling exists that progress is gradually being made.

PROBLEM WASTES

Problem wastes in Southern Africa originate from the following sources:

- Urban wastes. Because of high rainfall and sensitive aquifers in certain areas the control of leachates is essential. In addition, management in the rural areas is inadequate.
- Heavy metals in sewerage.
- Chemical wastes. In certain areas the control and handling of organic chemicals, pesticides, phosphogypsum, and other inorganics are problems.
- Coal discards, due to spontaneous ignition.

- Mining wastes. Gold mines are situated in the most populated areas of the country. Asbestos and other mines are situated in rural areas. These wastes have a tendency to pollute both the aquifer and the atmosphere.
- Effluents. These are defined as all notifiable wastes which are unacceptable in the sewerage system.

PROBLEM TECHNOLOGIES

The following problem technologies exist in Southern Africa:

- disposal of organic chemicals like PCBs and pesticides;
- management of co-disposal practices especially in high rainfall areas;
- deficient incineration practices;
- any waste management operation requiring water because of water shortage.

EXISTING AND CLOSED OR ABANDONED OPERATIONS

Abandoned mine dumps are located within the most populated areas of the country. They pollute both the surface and groundwater by increasing the salinity content.

Abandoned urban waste disposal sites are found to introduce leachate into the aquifer and the surface water system with detrimental effects. These sites are situated in areas with a sensitive aquifer like the Cape Flats and the Durban area.

Current practices of waste disposal sites in smaller local authorities tend to be very primitive. Table 2 indicates the current situation in South Africa.

TABLE 2
Summary of Current Waste Disposal Practices in South Africa (1)

Practice	Province				Total	Percentage
	Cape	Natal	OFS	TVL		
Uncontrolled disposal	121	27	46	31	225	27.6
Landfill	126	53	40	120	339	41.6
Burning	116	7	12	35	170	20.8
Composting	21	2	0	8	31	3.8
Recycling (partial)	0	4	7	14	25	3.1
No service	7	3	3	12	25	3.1
Total	391	96	108	220	815	100%

(1) Barnard (1983)

SUCCESSFUL PRACTICES

In Southern Africa waste management practices are landfill based. Successes can be identified in regard to landfills, chemical treatment plants, and the management of nuclear wastes.

The Vissershok landfill site in Cape Town is a class I co-disposal site that has been used for 15 years without the occasion of groundwater pollution. Linbro Park and Robinson Deep landfill sites are two examples of successful landfill sites in Johannesburg where the pollution levels are very low.

'Waste-Tech' operates successful chemical treatment plants in Durban, Cape Town and on the Witwatersrand. Also, nuclear waste is successfully disposed of by NUCOR, and recently a well-researched disposal facility has been commissioned.

PRIORITIES FOR NATIONAL ACTION

At least five priorities for national action exist. A survey must be done on the status of hazardous waste management on a regional basis for metropolitan and decentralized industrial area. Also, a data base must be set up on what materials are produced and where.

In other areas, the design of a Southern African legislative framework to enforce the proper handling of hazardous wastes is deemed essential. The licensing and control of the generation, transportation and disposal of hazardous wastes all perceived to be components of this system. A final priority is the design of proper treatment facilities for PCBs and other organic chemicals.

NATIONAL CONTROL SYSTEMS

Summary of legislation

The current control of hazardous waste is vested in various acts, ordinances and by-laws. All are intended to conserve the environment but the legislation is fragmented and administered by numerous agencies with detrimental results. The existing legislation covers the atmosphere, water, medicines, hazardous substances, sea dumping, nuclear installations and mining operations.

The list of acts is:

- The Water Act*, 1956
- The Health Act*, 1977
- The Hazardous Substances Act*, 1973
- The Regulation of Disposal of Pesticides Act*

---

* These are the principal acts.

- The Atmospheric Pollution Prevention Act, 1965
- The Medicines and Related Substances Control Act, 1965
- The Foodstuffs, Cosmetics and Disinfectants Act, 1972
- The Dumping at Sea Control Act, 1980
- The Mines and Works Act, 1956
- The Nuclear Installations Act, 1963
- The Precious Stones Act, 1964
- The Mining Rights Act, 1967
- The Atomic Energy Act, 1967
- The Agricultural Produce Export Act, 1971
- The Fertilizers, Farm Feeds, Agricultural Remedies and Stock Remedies Amendment Act, 1977
- The Sea Fisheries Act, 1973
- The Merchant Shipping Act, 1951
- The Prevention and Combating of Pollution of the Sea by Oil Act, 1971
- The Marine Traffic Act, 1981
- The Sea Shore Act, 1935
- Physical Planning Act, 1967

History of hazardous waste management

Since 1970 several major local authorities have attempted to deal with hazardous wastes. This resulted in the building of evaporation ponds for volume reduction and eventual solidification and later in co-disposal practices. Small incineration plants have been operative for many years.

The private sector entered the field in 1969 with the result that several companies are now active.

Since the discovery of gold in Johannesburg towards the end of the 19th century, the management of mining wastes has improved. In 1979 the Chamber of Mines produced "Guidelines for Environmental Protection". However, the environmental impact of mining operations remains a major concern.

Regulations for the licensing of disposal sites are currently being drafted in South Africa and will come into effect shortly. Until now, control was largely voluntary at a local level. The mechanism of control used to originate from the private sector only. However, since 1982 a mechanism was established with the institution of the Council for the Environment.

The Hazchem system will be extended in South Africa during 1985. This system will have reference to specific hazardous wastes. Other systems in operation are at local authority level, certain of which insist on documents giving proof of safe disposal. Private contractors have a delivery note system at point of uplift. The nuclear industry has a well-run control system in operation.

At present in the absence of national legislation, control is exercised largely by local government. Private companies organise their controls on a regional basis. With the implementation of the new political dispensation in South Africa, certain services will be provided on a regional basis. The control of disposal operations will then devolve to a regional level.

In South Africa indirect subsidies for the disposal of wastes are provided by local government, and, to a lesser extent, by central government in the form of low disposal charges. This is done mainly in the interests of economy and in cases where industry has neither the equipment nor the expertise.

In the Environment Conservation Act No. 100 of 1982 provisions was made for the co-ordination of all actions directed at or liable to have an influence on the environment; the establishment of a council for that purpose; the establishment by that council of committees in order to assist the council in the performance of its functions; the establishment of management committees in respect of certain nature areas; the appointment of honorary environment conservation officers; the making of certain regulations by the Minister; the assignment to local authorities of certain functions in terms of such regulations; the conferring upon the Minister, and the Administrator of a province, of certain powers in respect of certain local authorities; and to provide for incidental matters.

Systems have been developed in Southern Africa within the requirements of the current legal framework by local authorities, companies and the nuclear and mining industries. The Hazchem system has been introduced with specific reference to flammable liquids. This, however, will be expanded to include other materials in due course. Generally mining wastes, with certain exceptions, are transported by pipeline to containment sites.

In certain countries in Southern Africa the conservation ethic places the onus on the generator to take the necessary steps to protect the environment against the wastes that he produces. If the generator cannot cope, disposal companies or local authorities can be called upon.

In the transportation of wastes there is indirect and selective control. The transporter is held responsible for the safe transportation of hazardous wastes to a disposal facility.

At present it is the responsibility of local authorities and private companies to voluntarily ensure that the operation of facilities meets certain standards. The new regulations will be published in South Africa shortly.

In Southern Africa abandoned mine dumps and exhausted urban wastes sites fall under this category. Regarding the *urban waste sites*, follow-up action is being taken on a limited scale to monitor the continuing generation of leachates. In recent years many urban waste sites have been successfully rehabilitated by industry and local government, usually to produce public open space. This trend is increasing.

The *mine dumps* in South Africa are a constant source of pollution. The Chamber of Mines, however, has had outstanding success in the reclamation of these sites through vegetation programs and the control of water movement.

HAZARDOUS WASTE DEFINITION

In South Africa hazardous wastes are regarded as wastes which by the circumstances of use, quantity, concentration, or physical, chemical or infectious characteristics, may cause ill-health or may increase mortality to humans, fauna and flora, or may adversely affect the environment when improperly treated, stored, transported or disposed of, and exhibit the characteristics of ignitability, corrosivity, reactivity or EP toxicity.

The *purpose* of definitions based on the foregoing is to identify hazardous wastes and to trigger specialized storage, collection, and disposal, technologies. It would identify situations that would necessitate intervention by central government. This could either be by law enforcement, subsidization of corrective measures or by direct action.

The definition covers most hazardous wastes. However, the conscience of the people handling the waste is relied upon. Certain materials can be missed and dealt with retrospectively. The relationship is implicit in the defined effects the waste would have on public health, the water regime or the atmosphere.

AVAILABLE/SUMMARIZED INFORMATION

Urban wastes generally have a similar composition and generation rate to that in the rest of the world. The sources of industrial wastes are listed in Table 3 in decreasing order of volume generated.

TABLE 3
Industrial Sources and Relative Volumes of Hazardous Wastes (1)

Source	Relative volume (%)
1. Chemical, petroleum and coal product, rubber and plastic product manufacture	71
2. Fabricated metal product, machinery and equipment manufacture	12
3. Basic metal product, manufacture	10
4. Paper and paper product, manufacture, printing and publishing	4
5. Textile, wearing apparel and leather product manufature	2
6. Other	1

(1) van Rensburg (1983)

Mining wastes have a high potential for the release of hazardous components into the environment. These potential sources and relative volumes are listed in Table 4.

TABLE 4
Mining Activities in South Africa and Relative Production Quantities (1)

Mining activity	Production of waste (%)
Gold	41
Coal	29
Copper	8
Platinum	8
Limestone	7
Diamonds	6
Asbestos	1

(1) Malan (1983)

According to the 1980 census, the population of Southern Africa is 68 million. On the basis of a generation rate of 1 kg per day per person, 25 million tonnes of urban wastes are produced per annum. The addition of sewerage sludge significantly increases the quantities of hazardous wastes produced.

The major industries of Southern Africa are petroleum, chemical, foundry, and metal-finishing industries. The bulk of these industries are located in South Africa and are producing large amounts of wastes. Some of the quantities produced annually are listed in Table 5.

TABLE 5
Waste Produced by Major Industries in South Africa (1)

Waste type	Annual quantity tonnes ($\times 1000$)
Phosphogypsum	3000
Metallurgical slags	7000
Pulverised fuel ash	10,000

(1) von Rensburg (1983)

The quantities produced within the urban areas are more difficult to assess. For example, Johannesburg disposed of 3150 m^3 of hazardous wastes in 1979. In Cape Town, 14,000 tonnes of sludges and liquid industrial wastes and 60,000 tonnes of solid industrial wastes are deposited on landfills every year.

The relative monetary value of mining activities in Southern Africa are given in Table 6. South Africa is the center of the sector, followed by Zambia and Angola.

TABLE 6
Mineral Production in Southern Africa (1)

Country	Value of mineral production (R million/annum)
Botswana	22.1
Lesotho	0.3
SWA/Namibia	218.0
South Africa	3225.3
Swaziland	24.3
Angola	295.4
Malawi	0.1
Maputo	4.5
Zimbabwe	191.8

(1) Malan (1983)

The total production of minerals in South Africa is estimated at 254 megatonnes per annum. The accumulated production up to 1975, is estimated at 2858 megatonnes.

The kinds of wastes produced by the above-mentioned activities are typical of industrialized countries and include spent solvents, sludges, spent salts, tar, pitch, spent liquid, heavy metals, photographic chemicals, organic chemicals, oils and asbestos.

COLLECTION AND TRANSPORTATION SYSTEMS

The Hazchem system will be extended shortly in South Africa. At present this system controls the transport of pesticides, flammable liquids, and radioactive material. Indications are that this will be expanded to include other materials. The collection and transport of waste is largely controlled by local authorities and companies.

Manifest, or trip ticket, practices are undertaken by a few local authorities and waste management companies. Generally these permits gave the right of disposal as well as an indication of appropriate disposal technology. Transporters are controlled by a few local authorities through offensive trade permits. In Cape Town, the disposal permit includes certain controls on transportation. These systems are operated by the private and public sectors and a very few systems are run by large industries.

Special collection systems are used for some wastes. Waste oils are re-refined in South Africa. The resulting by-product, oil sludge, is disposed of together with oils in a fixation process using pulverized fuel ash. The final product is deposited on a class I containment site. Certain local authorities dispose of oil together with domestic wastes. One company is known to dispose of its waste oil via a pipeline into the sea.

Household hazardous wastes, comprising batteries, solvents and paints, are co-disposed on municipal disposal sites. Hospital wastes are handled, transported and incinerated, according to instructions laid down by the authorities. Abattoir wastes are treated and recycled to provide a valuable animal feed additive.

Ninety percent of the transportation capability is organized at local government level. However, companies are active throughout the country in handling hazardous wastes, some of which are transported to a containment site on the Witwatersrand.

TREATMENT, STORAGE AND DISPOSAL SYSTEMS

Unknown quantities of hazardous wastes are stored on industrial premises. Waste management companies also store some of these wastes on disposal sites prior to disposal. Radioactive wastes are generally stored in sealed containers. Short-lived isotopes are rendered harmless by storage whereas other more dangerous isotopes are transported to highly specialized disposal facilities.

Large scale incineration is not carried out in Southern Africa. In South Africa incineration is done on a small scale. These incinerators are owned and operated by local authorities, central government (habour installations), at hospitals, and by the private sector. Only specific wastes that cannot be dealt with safely on landfills, are incinerated. The incinerator temperatures generally range between 800°C–900°C and the heat produced is not utilized.

The disposal of wastes is carried out by uncontrolled dumping, by landfilling or by controlled sanitary landfilling. Over 50% of the local authorities in South Africa and most of those in the rest of Southern Africa simply dump their wastes which are then frequently burned (see Table 1). In the major centers in South Africa sanitary landfill is practised.

Three classes of sites have been identified for licensing purposes:

- Class I comprises containment sites with an *in situ* permeability of not greater than 10^{-6} cm per sec, which for all practical purposes is a sealed site.

- Class II sites are used for sanitary landfills and have an *in situ* permeability not greater than 10^{-3} and an attenuation zone of at least 2 m depth above the wet season water table.

- Class III sites are in direct hydraulic continuity with the groundwater and are unacceptable for disposal purposes unless the sites are adequately engineered.

Most of the major existing sites in South Africa are Class II sites. Some of the bigger local authorities and a waste disposal company operate Class I sites for the disposal of hazardous wastes. Toxic liquids are encapsulated before disposal in these sites. The cost of disposal is in the order of R1 per liter.

Surface impoundments such as pits, ponds and lagoons are fairly widely used by industry for the concentration and reduction of volumes. A few local authorities also use this system. The average unit cost of disposal is R2 per kiloliter neutralized effluent. Co-disposal is practised by a few local authorities and a waste disposal company. The unit cost of disposal is R5 per tonne.

Land farming, or mixing of wastes with soil, is not practised generally. The defence force uses spent-oils to lay the dust on military roads. Phosphogypsum and tannery sludge are being investigated as a soil conditioner.

Phosphogypsum is disposed into the sea 4 km off-shore via a pipeline by one company in South Africa which produces 6000 tonnes of phosphogypsum per day. A second pipeline is used to dispose of papermill effluent and flourine from a fertilizer company.

Many of the coastal local authorities use pipelines to dispose of sewerage into the sea. Waste oil is piped into the ocean after pre-treatment. Ammonia is pumped from a balancing tank into the ocean in one instance by a fertilizer company. Various industries along the coast are using pipelines to dispose of biodegradable wastes.

Recycling is widely practised in Southern Africa. These practices include:

10. SOUTHERN AFRICA

- Recycling of components in the urban waste stream, thereby effecting a 10% mass reduction. In some instances 54% of the packaging material is being recycled.
- The use of phosphogypsum as a soil conditioner.
- The use of metallurgical slags as rail ballast.
- The use of pulverized fuel ash in oil fixation processes and as a building material.
- The use of chicken litter and selected sewage sludge, after irradiation, as an animal feed.
- The recycling of lead, cadmium and tin.
- The re-use of old tires for tarring road surfaces.
- The use of buffing dust, rubber off-cuts and plastic wastes to produce various goods such as dustbins.
- The recycling of mineral oils is widely practised and encouraged.

Programs and activities to reduce waste generation include the following:

- Industry recycles in-house for economic reasons.
- Current research is under way to establish the limits of urban recycling with a view to reducing the waste stream.
- A well-developed low technology industry exists in Southern Africa for the recycling of glass, paper, plastics, cardboard and metal.

Most of the disposal sites in Southern Africa are owned by local authorities. A few are owned by waste disposal companies. A few industries have their own sites which are used for storage and/or disposal. There is some movement of waste from developed states to lesser developed states where disposal costs are lower. Quantities are unknown.

ASSESSMENT

The control of hazardous wastes is carried out at the local level on a fragmented basis. Enforcement from central government is only practised where hazardous wastes affect the water regime or the atmosphere. In these cases control actions are enforced through the inspectorates of both the Departments of Health and Water Affairs.

The Hazchem system is being extended to other hazardous wastes. The licensing of disposal facilities will continue to take effect.

There is no co-ordinated control. Existing controls are fragmented and localized. Furthermore, the current controls are reactive and not proactive.

It is expected that the proposed legislation on the licensing of landfill sites will improve the standards and facilitate improved control. The newly established Council for the Environment with its sub-committees will create a forum from which it is hoped that better control and protection of the environment will result.

The following hazardous wastes management practices and activities are examples of what is being done well:

- The establishment of the Council for the Environment with its sub-committees, (e.g. Working Group for Solid Wastes and Recycling) is a move forward in the protection of the environment.

- At the Council for Scientific and Industrial Research (CSIR) the Solid Waste Management Section is undertaking goal oriented research projects, which are designed to solve problems regarding urban, industrial, mining and organic wastes.

- The Institute of Waste Management (IWM) is organized on a national and a regional basis. The IWM runs conferences to bring its members up-to-date on developments in the field of solid waste both in South Africa and abroad. In addition it runs a formal educational programme.

- The IWM is acting as a regional catalyst in the field of solid waste management.

- Certain local authorities and the largest waste disposal company are setting very high standards for the disposal of hazardous wastes.

- In recent years mining and industry have expended large amounts of money to rehabilitate mine waste dumps and open-cast mines.

- The Hazchem system has been implemented.

The following aspects can be regarded as being done poorly:

- A well-informed body of public opinion regarding hazardous wastes has not yet been formed.

- A quantitative data base on hazardous wastes that inhibits adequate control does not exist.

- The absence of specific national legislation.

- The controls on the transportation and disposal of hazardous wastes are incomplete and inadequate.

The control of hazardous waste management via a national legislative framework must be instituted on the basis of a current data base on hazardous and potentially hazardous wastes, if improved practices are to occur. The transfer of applied technology from the more developed subregions to the lesser developed sub-regions is of the utmost importance in Southern Africa. In certain centers in South Africa these technologies are known. However, means must be found to transfer this information and the associated skills to where they are needed.

CONCLUSION

Throughout Southern Africa there is a direct link between the degree of industrial development and the quality of waste management. The highest industrial concentrations are found in the more developed areas and thus there is some voluntary and statutory control of hazardous wastes. However, in order to be able to conserve the environment, steps must be taken now to prevent practices which will create potentially serious environmental degradation throughout the region.

BIBLIOGRAPHY

Barnard. J. J. (1983). *Proposed legislation in connection with control of waste disposal sites*. Proceedings of the Seminar on the Handling of Urban Wastes on a Regional Basis. CSIR, Pretoria.

Humayun, G. (1984). A socio-economic digest of the Third World. *South — The Third World Magazine*, October.

Jorgensen, A. A. (1983). Rail transport in Southern Africa. *Africa Insight* 13(1).

Malan, J. J. (1983). *The Role of the Waste Management Section at the CSIR*. Proceedings of the Seminar on the Behaviour of Radium and Rodon in the Environment. NUCOR, Pretoria.

Malan, T. (1983). Southern Africa: notes on resources and structures. *Africa Insight* 13(1).

Van Rensburg, B. (1982). *A study of hazardous waste production and disposal in South Africa report*.

Van Rensburg, B. (1983). *An objective review of the present situation regarding solid waste generation, recovery and disposal in South Africa report*.

Verrier, S. J. (1982). *The choice of appropriate technologies in Southern Africa*.

Zietsman, H. L. (1981). *Economic Atlas of South Africa*. Institute of Cartographic Analysis, University of Stellenbosch.

11

Hazardous Waste Management in SPAIN and PORTUGAL

JULIAN URIARTE JAUREGUIXAR

Secretary General, ATEGRUS, Bilbao, Spain

OVERVIEW

In Spain there is still no official legislation about hazardous waste. This means that it is quite impossible to determinate the quantities generated and the composition of what is disposed of because most of it is done without the knowledge of the authorities. This lack of strict legislation allows for reckless disposal of hazardous waste leading to severe damage to the environment. However, we are very concerned with the problem of environmental pollution, and hard work has improved the present situation.

The industrial waste is treated in the same way as urban waste. Only a few industries, producing more dangerous waste, take greater safety precautions.

The situation in Portugal is more or less the same.

NATIONAL CONTROL SYSTEM

The only act which deals with waste dates from 1975. It is a general law about urban waste and hazardous waste is not taken into account.

At present, legislation about hazardous industrial waste has been developed, and is waiting to be approved. Because of the political situation it is difficult to say when the act will be finalized.

In Portugal matters are even worse. No steps have been taken to prepare such a law.

On a regional basis in Spain, Cataluña passed an act in 1983 with regulations added in 1984-85. A type of trip ticket is needed in Cataluña to transport hazardous industrial waste. Waste is most often transported by truck because of the relatively small quantities involved.

With no legislation, there are no subsidies, no responsibilities, no programs for dealing with old, closed or abandoned sites.

DEFINITION, SOURCES AND QUANTITIES OF HAZARDOUS WASTE

With no legislation there is obviously no legal definition of hazardous waste. In Spain, data stating source, quantity and type of hazardous waste will not be available until the act is passed. Even then it will be difficult to determinate these data because the tonnage of the untreated waste will be very large compared with what will be treated.

There are statistics developed by regional and national organizations but these are not very reliable because of the small amount treated. The 1975 law excludes mining waste. The Catalan act also excludes mining and hospital wastes, but not those from laboratories.

In Portugal, the data obtained are not very reliable but the information based on a survey of Portuguese industry is given in Table 1. This is based on a survey made by the ANRED of French industry during 1973-80. Data were developed using two criteria, the numbers of workers and the number of products.

Analysing the results according to the two criteria, a concordance between the two factors can be seen. In some activities there is a great difference. This can be caused by the different nature of the units in which the statistics are expressed.

COLLECTION AND TRANSPORTATION

In Spain and Portugal there are established regulations for transport by road adapted to international regulations. In both countries the regulation is for hazardous goods but not specifically for waste.

TREATMENT, STORAGE AND DISPOSAL SYSTEMS

TABLE 1
Estimation of the Annual Production of Hazardous Wastes in Portugal

	According to the number of workers (thousand tonnes per year)	According to the number of products (thousand tonnes per year)
Food and agricultural industry	231	168
Fuels and energy	65	30
Extractive and metal industry	209	1029
Glass and building materials	110	384
Chemicals	134	133
Electrical and mechanical industry	135	52
Paper and coal industry	1.8	2.4
Shoe and clothes industry	1928	658
Wood industry	2.9	14
Graphic arts and publishing	0.7	0.4
	2817.4	2470.8

Incineration systems

In Spain there is no centralized facility for incineration. There are only a few firms, like Flix or Cepsa, which have incinerators for their own use. In general, petrochemical manufacturers and paper factories incinerate for their own use with steam recovery. It is very difficult to know the quantities incinerated, although some plants are being studied.

Landfills and disposal of waste on the land

In most cases industrial waste is disposed of in urban landfills without any control. The Catalan Act forbids this mode of action. The Spanish draft also bans it.

The only known underground mine disposal system is the plaster mine at Hindronor. Surface impoundments such as ponds and lagoons are used by some chemical and petrochemical firms. Co-disposal with household garbage is performed at most landfill sites. It occurs because there is no legislation to regulate it. Landfarming also exists without any control.

There is not a centralized landfill for hazardous industrial waste in Spain. Only one company, Pesa, is authorized to use a landfill for its own waste (chromite manufacture).

There is a landfill (Fontsanta, near Barcelona) for inert industrial waste with a volume of 650 tonnes per day.

Physical and chemical treatment

There is only one firm in existence (Hidronor) with two facilities in Guipúzcoa and Barcelona. In Guipúzcoa there is oxidation and reduction, neutralization, vacuum filtration, settling of the final effluent and storage of the dry cake in an appropriately conditioned plaster mine.

In Barcelona there is oxidation and reduction, neutralization, solidification and disposal in a sanitary landfill. The volume treated in the facility of Guipúzcoa is 50-60,000 tonnes per year. In Barcelona the volume is 1000 tonnes per year. Physicochemical treatments closely depend on the nature of the waste to be disposed of, and by extension the manufacturing activity.

Energy recovery and recycling

The most common recycling processes are the following:

- Use of the waste as a raw material for another process in the same firm which generates it — for example lime oxide and pulp waste used in the manufacture of paper.
- Waste treatment for its later use as a raw material in another process.
- Waste reconditioning for its later use as a raw material in the same or similar process in which it was generated. For example: used oil recovery, and solvent distillation. The treated oil is used as an energy product. Solvents are recycled at in-house industrial recycling facilities.
- Waste is used as primary or secondary fuel, with steam and electricity recovery. An example in waste liquor incineration in paper making.

Description of programs to reduce waste generation

At present there are no definite programs. Planning has been done to encourage the use of clean technology but there is scarce information about it.

In some large private companies there are programs for their own benefit. There are no public initiatives. The Federation of Chemical Industries is considering this.

Imports and exports

Few data on exports are available. The quantity involved is probably not significant although the effect on environmental quality is important. Nearly 1000 tonnes per year are exported. Hidronor transports 800 tonnes per year, basically pesticides, chlorinate and pharmacy products.

Importing waste is a great problem for Spain and the country is becoming a landfill for other countries.

In Portugal all transport of waste abroad is clandestine and secret.

ASSESSMENT AND FUTURE DIRECTION

In Spain the percentage of controlled disposal of waste is around 5–10%. With our entrance into the EEC in January, 1986, we expect to improve our present situation.

12

Hazardous Waste Management in SWEDEN

KENNETH ANDERSSON

AB Graab-Kemi, Gothenburg, Sweden

OVERVIEW

Since the middle of the 1970s definite efforts have been made in Sweden to deal with hazardous waste by methods compatible with environmental interests. Overall, legislative directives were reviewed and supplemented with a special ordinance. At that time there were already a number of smaller plants for the destruction or recycling of hazardous waste. In 1983, SAKAB's plant in Norrtorp was ready to be taken into service. This is the central plant for hazardous waste in Sweden. Apart from the 96%-owned SAKAB treatment corporation, there are a number of smaller plants approved by law for the destruction, processing or recycling of hazardous wastes, often in conjunction with other industrial facilities.

However, in spite of the various facilities in Sweden for dealing with hazardous waste, not all the problems have been solved. The wastes that present problems, and for which there are still limited facilities for final disposal, are: PCB-contaminated waste; organic wastes with high chlorine content; and mercury-contaminated waste.

One technical problem which needs attention and which is currently being studied in Sweden is the handling and emptying of oils containing PCB from transformers and capacitors. Another difficult problem is the cleaning of flue gas when incinerating organic waste with high chlorine content. The latter problem is of such dimensions that international cooperation is a must if technical and economical solutions to the problem are to be found.

Sites which have been closed are mainly landfills used for the disposal of hazardous waste. These are either purely industrial landfills or landfills for co-disposal with other non-hazardous waste. There were also one or two incineration plants — in Stockholm (*Industridestillation*) and Stenungsund, for example. Both are now closed.

Successful practices for the treatment or recycling of hazardous waste are to be found at SAKAB and the 20 or so smaller plants. A number of different systems have been established in Sweden for the collection and transportation of hazardous waste.

Most of the resources necessary for the treatment and disposal of hazardous waste do exist in Sweden. The measures now necessary if complete control is to be gained over hazardous waste, and which are currently receiving priority are:

- increased resources in the form of regional reception stations and temporary storage,
- more efficient supervision of the correct handling and disposal of hazardous waste.

NATIONAL CONTROL SYSTEMS

Summary of legislation

The treatment of hazardous waste in Sweden is regulated by a number of laws, ordinances and regulations at various levels. In older legislation the concept of hazardous waste did not exist.

It was not until 1975, with the "Ordinance on Hazardous Waste" that a comprehensive concept of such waste—hazardous waste—was introduced.

This ordinance regulates the handling of such waste, and covers the following points:

- definition (categorization) of hazardous waste,
- obligation for anyone whose commercial activities involve the generation of such waste to provide statements concerning the type, composition, amount, and treatment of hazardous waste as and when required by the National Environment Protection Board,
- transportation of hazardous waste,
- final disposal/treatment of the waste,
- export of hazardous waste,
- import of hazardous waste,
- supervision, and observance of the ordinance.

The ordinance does not apply to hazardous wastes included in household waste. Apart from the above-mentioned ordinance there are a number of laws, official announcements etc. relating to hazardous waste in Sweden.

History of hazardous waste management

Although many of the laws and ordinances concerning hazardous waste are quite old, it took a relatively long time for the state and local authorities to make any serious attempt to bring the waste situation under control.

The Municipality of Gothenburg — the second largest city in Sweden and the country's largest port—was the first in the country to introduce a monopoly on the collection and management of hazardous waste. This took place in 1973. With the introduction of this monopoly, limited resources were also established in the form of regional reception and temporary storage stations, along with certain facilities for treatment of waste for Gothenburg and the surrounding municipalities.

SAKAB was established in its present form in 1974, and is responsible for the necessary resources for the treatment of hazardous waste on a national basis, in accordance with the Ordinance on Hazardous Waste which at that time was being circulated for consideration by the bodies concerned.

There were at that time already a number of companies treating certain types of hazardous waste. These were primarily handling oil spills, solvents and lubricating oils, and were mainly occupied with recycling.

From January 1, 1986, the municipalities have had to take care of the collection and transportation themselves, and are now establishing temporary storage facilities, and in some cases also plants for pretreatment.

Licenses and manifest system

Requirements for permits for the construction and operation of plants for final disposal of hazardous waste are contained in the Environment Protection Act, and the Ordinance on Hazardous Waste.

The only corporation which does not need to apply to the government for permission is SAKAB. However, permission pursuant to the Environment Protection Act is also required for SAKAB.

Through these basic permits all the requirements of other relevant legislation must be satisfied.

All transportation of hazardous waste is subject to current regulations for the form of transport in question. For example, in the case of land transportation the vehicle must carry a plate which states the primary risk and classification of the load, as well as documents which state the chemical composition etc., of the goods — a complete declaration of the load, in other words. If hazardous waste is to be transported out of the country, a special permit is required. The transporter must be authorized by the County Administration.

Permission for the construction and operation of a plant for the management of hazardous waste is required pursuant to the Environment Protection Act and, if final disposal is involved, also from the government. Permission under the Environment Protection Act may be granted only by a special licensing board for environmental protection, the Franchise Board of Environment Protection, after extensive investigation. Appeal against the decision of the licensing board may be made to the government.

The National Swedish Environment Protection Board has the overall national responsibility for supervising the observance both of conditions laid down by the licensing board and of any requirements imposed by other relevant legislation. Regionally, the County Administration has the same responsibility, as do environment and health protection boards at the local level in each municipality.

Subsidies

The principle in Sweden is that the cost of hazardous waste management shall be borne in total by the generator of the waste. It is also desirable that each type of waste carries its own costs.

In spite of these general principles, state subsidies have — to a certain limited extent—been provided for hazardous waste management, particularly during the 1970s, in the form of investment subsidies for certain treatment plants, and grants to cover costs in connection with the long-term storage of hazardous waste.

Transport systems

The great majority of hazardous waste in Sweden is transported by road, though some is moved by rail. In future, rail transportation will increase since SAKAB's rail reception facilities have now been established.

Liquid waste is transported in tank trucks and in sludge suction trucks, or by rail tank transports. Other waste is normally transported in barrels, standard containers or environmental boxes, for example. Restrictions on mixed loads must be observed.

All transports must be authorized, and the necessary transport documents must accompany the transport. The hazardous waste is frequently kept in temporary storage until sufficient quantities have been accumulated for transportation or treatment to be economical.

Responsibilities

The generator shall, on request, render an account of the type, quantity and composition of hazardous waste generated as a result of commercial activities. On transfer or delivery, a declaration of the waste is to be provided on a special form.

The transporter shall be authorized, and all necessary documents shall be carried and plates displayed during transport. Treatment facilities shall have the necessary permits required by current legislation.

Program for dealing with old, closed or abandoned sites

All active companies in Sweden which at present treat hazardous waste possess the necessary permits. Facilities which have closed or been forbidden have been mapped out and in some cases decontaminated.

At present a nationwide inventory of old landfills is being carried out by the supervisory authority (the National Swedish Environment Protection Board, in conjunction with the Swedish Association of Local Authorities) to find out which landfills also contain hazardous waste. Landfills are categorized according to decontamination requirements.

DEFINITIONS OF HAZARDOUS WASTE

A guideline list of hazardous waste is published by the National Swedish Environment Protection Board. This list constitutes the basis for application of the Ordinance on Hazardous Waste. At present the list contains 12 main groups of wastes that are defined as hazardous wastes (see Table 1). The aim of this list is to define the types of waste classed as hazardous, and thereby to indicate that the handling of such waste requires special permission by law. The main objective is to enable control to be made so that hazardous waste is dealt with in the correct manner.

The main problem as far as supervision and control is concerned is the lack of personnel in the supervisory authorities. This makes efficient supervision and control difficult, in spite of the extremely efficient systems which have been worked out in certain parts of the country as a result of extra efforts during the last few years. These special supervision systems have led to considerably larger amounts of hazardous waste being dealt with in an environmentally satisfactory manner. Cooperation between the supervisory authorities and waste-generating industries has also been positive.

TABLE 1
Guideline List of Hazardous Waste

Main categories
(1) Oil waste
(2) Solvent waste
(3) Paint and varnish waste
(4) Adhesive waste
(5) Acid or alkali waste
(6) Waste containing cadmium
(7) Waste containing mercury
(8) Waste containing compounds of antimony, arsenic, barium, beryllium, lead, cobalt, copper, chrome, nickel, selenium, silver, thallium, tin or zinc.
(9) Waste containing cyanide
(10) Waste containing PCB
(11) Pesticide waste
(12) Laboratory waste

The term hazardous waste refers in this list to materials, raw materials, products, intermediate products, by-products or chemicals which have been polluted, destroyed, prohibited, or cannot for any other reason be used for their originally intended purpose.

The concept of hazardous waste as defined in the 12 main groups of Table 1 is relatively comprehensive. There are, however, certain deficiencies — mainly concerning the definition of ordinary industrial waste, and rules for handling it.

Now that all local authorities in Sweden have introduced monopolies on the collection of hazardous waste, only non-hazardous industrial waste, construction waste and fill can be managed with little or no control. Permission is, however, required from the County Administration for landfills intended for this sort of waste. Hazardous waste is, in principle, not treated together with other types of waste.

SOURCES AND QUANTITIES OF HAZARDOUS WASTE

The National Swedish Environment Protection Board was commissioned by the government to carry out a study of hazardous waste in this country in 1980. In the project "Hazardous Waste 1980" studies of the properties of waste and the methods of dealing with it were included. A questionnaire survey was carried out in which almost 30,000 different types of workplace were selected in consultation with local councils and County Administrations. When the project "Hazardous Waste 1980" was implemented, hazardous waste was defined under 13 main groups. This has now been changed to 12 main groups, as shown in Table 1.

The following tables have been drawn from the above-mentioned project:

- total amounts of hazardous waste in tonnes for 1980, for the entire country, according to the 13 main waste categories (Table 2),
- quantities of hazardous waste for 1980, percentage distribution among main industrial branches (Table 3),
- composition of hazardous waste, percentage distribution by main component categories (Table 4),
- consistency of hazardous waste, percentage distribution according to solid, sludge and liquid (Table 5).

TABLE 2
Quantities of Hazardous Waste in Sweden

Category	Quantity 1980 (tonnes)
(1) Oil waste	183,000
(2) Solvent waste	34,600
(3) Paint waste	20,700
(4) Adhesive waste	10,800
(5) Acid or alkali waste	72,300
(6) Cadmium waste	480
(7) Mercury waste	440
(8) Other metal wastes	120,000
(9) Cyanide waste	3500
(10) PCB waste	31
(11) Pesticide waste	580
(12) Other chemical wastes	40,700
(13) Laboratory waste	300
Total (tonnes)	488,200

COLLECTION AND TRANSPORTATION SYSTEMS

Since all local authorities in Sweden have now introduced monopolies on the collection and transportation of hazardous waste, control and supervision in this area will primarily be the responsibility of the relevant department of the local authority, but will also be carried out by the local environment and health protection board and the regional County Administration.

TABLE 3
Quantities of Hazardous Waste 1980, by Industry

Waste category	Forestry/agriculture (%)	Chemical industry (%)	Engineering (%)	Iron/steel etc. (%)	Other industries (%)	Other activities (%)
(1) Oil waste	5	15	23	14	5	38
(2) Solvent waste		52	16	1	8	23
(3) Paint waste		11	49	2	27	11
(4) Adhesive waste		5			50	45
(5) Acid or alkali waste		34	35	19	6	6
(6) Cadmium waste			57	41	1	1
(7) Mercury waste		86	2		2	10
(8) Other metal wastes		1	36	55	5	3
(9) Cyanide waste			66	1		33
(10) PCB waste	3	1	8	1	1	86
(11) Pesticide waste	27	13			51	9
(12) Other chemical wastes		77	10	4	6	3
(13) Laboratory waste		11	3	9	19	58
Total (%)	2	21	28	22	8	19

TABLE 4
Composition of Hazardous Waste by Main Component Category

Waste category	Poisonous/corrosive substances including acids and heavy metals (%)	Solvents (%)	Oil (%)	Paint, varnish, glue (%)	Water (%)	Solid non-hazardous components (%)
(1) Oil waste	1	1	46	1	33	20
(2) Solvent waste	1	64	7		21	6
(3) Paint waste	1	18	1	32	36	12
(4) Adhesive waste	3	1		9	79	8
(5) Acid or alkali waste	13		5		76	6
(6) Cadmium waste	20				1	79
(7) Mercury waste	0.02				1	99
(8) Other metal wastes	16				48	36
(9) Cyanide waste	4				86	10
(10) PCB waste	8		6	30		56
(11) Pesticide waste	23	13	6		33	25
(12) Other chemical wastes	21	12	3		32	32
(13) Laboratory waste	13	44	5	1	27	10
Total (%)	8	6	20	1	41	24

TABLE 5
Consistency of Hazardous Waste

Waste category	Solid (%)	Sludge (%)	Liquid (%)
(1) Oil waste	14	14	72
(2) Solvent waste	1	5	94
(3) Paint and varnish waste	21	48	31
(4) Adhesive waste	13	17	70
(5) Acid or alkali waste	3	8	89
(6) Cadmium waste	70	29	1
(7) Mercury waste	2	98	—
(8) Other metal wastes	57	22	21
(9) Cyanide waste	4	7	89
(10) PCB waste	85	—	15
(11) Pesticide waste	41	30	29
(12) Other chemical wastes	14	6	80
(13) Laboratory waste	2	2	96
Total (%)	22	16	62

Both SAKAB and the major collection organizations in Sweden have produced their own transport and handling documents in the form of waste declaration forms giving various details such as generator, type, composition and quantity of waste, ADR codes, risk and safety markings etc.

All serious waste management in Sweden is documented on a waste declaration form before treatment commences. The declaration then follows the waste both through the administration and, in practice, through all stages of the waste management process — collection, transportation, temporary storage and final treatment or recycling. The waste declaration can in most cases also serve as a formal transport document in accordance with current ADR rules.

Work is at present under way in this country, led by the National Swedish Environment Protection Board, with the aim of developing a computerized information and control system for hazardous waste. Existing declaration systems in modified form are here seen as the basic elements of a nationwide control system which would arise more or less automatically, since waste declarations are anyway necessary both for practical management of the waste and for fulfilment of ADR transport rules.

Local councils will be required by law (see above) to collect hazardous waste. In such cases as the local council itself does not carry out the work, a contractor may be engaged. Such contractors must be licensed by the County Administration.

For transports outside the local council's area to a reception and

TABLE 6
Methods of Transport for Hazardous Waste by Percentage Distribution

Waste category	Bulk/tank (%)	Case-goods Total (%)	Barrels steel (%)	Barrels plastic (%)	Plastic canisters (%)	Other (%)
(1) Oils	88	12	94	0	6	0
(2) Solvents	82	18	98	0	1	1
(3) Paints	75	25	96	0	2	2
(4) Adhesives	83	17	86	8	1	5
(5) Acid/alkali	96	4	50	2	15	33
(6) Cadmium	41	59	98	0	1	1
(7) Mercury	73	27	23	0	64	13
(8) Other metals	91	9	66	2	23	9
(9) Cyanide	13	87	89	6	4	1
(10) PCB	1	99	97	0	0	3
(11) Pesticides	68	32	86	4	0	10
(12) Other chemicals	85	15	68	3	26	3
(13) Laboratory	25	75	75	0	14	11
Total (%)	88	12	86	2	8	4

temporary storage station situated elsewhere, for example, a licensed transporter is also required. Transportation may be carried out both by public (local council) or by private transporters. Private transporters are required to be licensed. For a detailed look at hazardous waste transport methods, see Table 6.

Waste oil collection systems

Waste oil is normally collected selectively according to the following categories: fuel oil; lubricating oil; oil-based emulsions; and oily sludge.

Waste oil is collected from the major generators by tank truck. Smaller quantities of oil may, however, also be collected by tank truck, but in such cases a number of generators are visited for the sake of transport economy. Collection and transportation of very small amounts of waste oil may also take place in barrels.

Pretreated waste oil — after removal of water — is also transported in smaller coastal tankers to refineries.

As has been mentioned above, the aim is for each type of waste to carry its own costs. This approach results in a not inconsiderable reimbursement for the generator after deduction for collection and treatment (recycling), particularly in the case of better quality waste oil and lubricating oil.

Household hazardous wastes collection systems

In certain municipalities a small truck drives around the various areas, according to a schedule, and collects household hazardous waste. The general public therefore have the opportunity of disposing of their hazardous waste where they live.

Recently, several municipalities have started collection of batteries containing mercury and cadmium in small boxes resembling post boxes, in which the batteries may be left. Batteries may also be left at shops where they are sold — photography shops, toy shops, opticians, etc. Old unused medicines and broken thermometers may be handed in at chemists, who then send them on.

The general public may also, in certain municipalities, book collection of hazardous waste at the next collection of ordinary household waste. This must be well packaged and clearly marked. This procedure is followed in the countryside. All collection of hazardous waste from households in Sweden is carried out free of charge to the general public.

Hazardous waste other than waste oil and household waste — mainly from industry, workshops, laboratories etc, are most often collected under the auspices of the local council. Once the waste has been declared, it is collected, sorted and classified, and then either kept in temporary storage or sent directly for recycling or final treatment.

Only in very large municipalities do the conditions exist for establishing a local organization with the necessary knowledge and expertise. Gothenburg may be given as an example — the municipally-owned company Graab-Kemi works not only for the municipality of Gothenburg, but also has contracts for the collection of hazardous waste in about 20 of the surrounding municipalities. In this way, a municipal resource has become a regional one. There are a number of organizations similar to the one in Gothenburg, particularly in the large city areas.

Other systems also exist for the organized collection of hazardous waste. Some private contractors have become specialized in hazardous waste, and one or two major waste contractors dealing with household waste on a nationwide basis now also offer a service for the collection of hazardous waste. These contractors are in a position to maintain the necessary competence either because of their specialization or because of their size.

There is no national organization for the collection of hazardous waste in Sweden. The state-owned waste management corporation SAKAB has, however, a number of regional reception stations in different parts of the country, to which municipalities can transport hazardous waste for onward transport by SAKAB to its own plant in Norrtorp.

As has been mentioned, municipal monopolies for the collection and transportation of hazardous waste have recently been introduced throughout Sweden.

TREATMENT, STORAGE AND DISPOSAL SYSTEMS

A comprehensive picture of the management of hazardous waste in Sweden during 1980 is provided by Table 7, both according to distribution by

TABLE 7
Management of Hazardous Waste in Sweden in 1980

Method	
Incineration	20%
Recycling/conversion	24%
Other use	2%
Landfill	40%
Sewerage	12%
Other methods	1%
Storage (management not yet solved)	1%
	100%
Manager	
Own management by generator (industries etc.)	36%
SAKAB	10%
Other disposal companies	33%
Municipalities	21%
	100%

percent of different waste methods and according to waste managers. Distribution by percent has, however, changed somewhat since SAKAB's facilities in Norrtorp started operating during 1983. Both for incineration and for SAKAB, the proportion of the total amount of hazardous waste is somewhat higher.

Storage systems

Storage before final disposal takes place at two different levels — temporary storage of hazardous waste at reception stations around the country, and long-term storage of certain types of waste.

Temporary storage is done for two main reasons: to await sufficient transportation and/or treatment quantities, or as buffer storage when the treatment plant is running at full capacity. Long-term storage is also performed for reasons other than a temporary shortage of treatment capacity — when adequate treatment resources do not exist.

Storage of hazardous waste is carried out in special buildings which have no drains and which are divided into various cells according to primary risk factor, e.g. poisonous, corrosive, flammable, self-igniting etc.

Liquid waste in large quantities is stored in cisterns which may be either heated or unheated. Hazardous waste containing volatile substances are stored in walled-in cisterns which often have floating lids to reduce emissions.

Storage sites and temporary storage facilities are strategically placed throughout the land in a way which best corresponds to the need. These stores may be administered by municipalities, by SAKAB, or by private companies. SAKAB is also responsible for the long-term storage necessary in the case of inadequate treatment resources.

Incineration systems

Land incineration is carried out in SAKAB's plant and at a very few industrial plants for internally-generated waste, and also in an incinerator for medical waste owned jointly by various Swedish pharmaceutical companies.

The main plant is, as has been mentioned, SAKAB's rotary kiln in Norrtorp, which has a capacity of 33,000 tonnes per year. The plant began operating in 1983. The types of waste which SAKAB accepts for incineration are primarily oil waste, solvent waste, paint and varnish waste, adhesive waste, waste containing PCBs, pesticide waste and certain other types of hazardous chemical remains.

For a comprehensive picture of SAKAB's resources for the treatment of hazardous waste, see Table 8.

TABLE 8
SAKAB, at Norrtorp, is the Central Plant for the Treatment of Hazardous Waste in Sweden

SAKAB expects to be able to accept the following types of hazardous waste in the quantities shown:

Waste category	Tonnes per year
Oil waste	30,000
Solvent waste	7000
Paint or varnish waste	7000
Adhesive waste	500
Acidic or alkaline waste	3000
Waste containing cadmium	100
Waste containing mercury	1000
Waste contaminated by substances containing heavy metals	5000
Waste containing cyanide	200
Waste containing PCBs	100
Waste containing pesticides	200
Hazardous chemical remains, by-products, etc., that cannot be included under other main headings	5000
Laboratory waste	50
Other	1000
Total approx.	60,000

Sweden has no resources for sea incineration. If this is necessary then the waste must be exported. Sweden has not exported any waste for sea incineration since January 1984.

Landfills and disposal of waste on the land

Landfills exist in Sweden for treated hazardous waste, such as metal hydroxide sludge and other heavy metal waste. Such landfills conform to strict regulations. They are often sealed in reinforced plastic, treated with lime before sealing, and covered well. These landfills have separate and well-documented sections for each type of waste, to enable controls to be carried out in the future. Landfills are also used to some extent for paint and adhesive waste, as well as for oily sludge.

Underground disposal in mines or caves is not used in Sweden, nor are underground injections in deep wells.

Surface impoundments such as pits, ponds and lagoons are used only in exceptional cases in Sweden. These may be lagoons for separation of oil spill or for certain types of industrial sludge which are converted to non-hazardous waste after decantation, and then treated in sewage treatment plants.

Co-disposal with refuse, construction waste or other non-hazardous materials is not normally undertaken in Sweden. Land farms or disposal of waste mixed with soils is also not undertaken in Sweden.

Ocean disposal is not used in Sweden for hazardous waste. Swedish legislation and international agreements ratified by Sweden concerning discharge at sea or dumping at sea forbid this form of disposal.

Energy recovery and recycling

Those types of hazardous wastes which are mainly recycled in Sweden are oil waste, lubricating oil, solvents and silver waste.

Oil waste is recycled by some treatment facilities. The treated oil is used as an energy product by being sold to industries or other concerns, which have permits to burn this type of oil and therefore possess approved flue gas cleaning facilities. Lubricating oil is re-refined at a plant for oil production, and reused in further production of lubricating oil. Solvents are recycled by distillation at some plants, both at commercial treatment plants and in internal industrial recycling facilities. Silver is recycled from photographic waste at some plants. Mercury is recovered at certain facilities out of wastes from dental surgeries. A relatively new method for recycling mercury is also being installed at the SAKAB plant. The method works by heating the mercury-contaminated waste until the mercury evaporates, after which it is condensed and recycled as mercury. Mercury waste with certain organic impurities can also be treated in after-burners. Heavy metals are recycled from wastes from the iron and steel industries, and also from surface treatment waste, in a recently built plasma reactor.

At SAKAB's plant in Norrtorp, energy recovery takes place by sending surplus heat from incineration in the rotary kiln via a heat exchanger to the district heating network of the nearby town for household heating and other uses.

Description of programs to reduce waste generation

The general attitude in Sweden to a reduction in the quantity of waste is that recycling should take place at source and that attempts should be made to replace those substances which give rise to hazardous waste in different production processes with non-hazardous substances.

A good example of the latter is the move from solvent-based paints to water-based paints in many businesses. This positive trend has been speeded up not only by the general desire to reduce quantities of hazardous waste and demands for improved working environments, but also by rising prices for solvents.

Another example is the replacement of mercury batteries by batteries containing less mercury. According to an action program adopted by the Swedish government in 1985, mercury from batteries in household waste is to be reduced by 75% during the years up to 1987.

Recycling of various chemicals has for some time been carried out at the source of production in many industrial processes in Sweden.

Operation and ownership of facilities

Some disposal facilities in Sweden are wholly in private ownership, and some are owned jointly by state and private interests. SAKAB, for example is 96% state-owned, with the Swedish Association of Local Authorities and the Federation of Swedish Industries owning 2% each.

Imports and exports

The great majority of hazardous waste, about 97%, is disposed of inside the country. Some of the remaining 3% is put into long-term storage and some is exported to other countries. About 10-15,000 tonnes a year of hazardous waste are exported. The treatment of the waste in the receiving countries consists mostly of some kind of recycling.

ASSESSMENT

Legislation concerning hazardous waste in Sweden requires any person who intends to handle such waste to apply for permission for such activities. Any person whose commercial operations lead to the generation of hazardous waste must render an account of the type, quantity and composition of the waste, on request by the National Swedish Environment Protection Board.

Supervision and control of operations involving hazardous waste come under the authority of the National Swedish Environment Protection Board at a national level, the County Administration at the regional level and environmental and health protection boards at the local level. Staff resources are too limited at all levels. Routine supervision is impossible, except at certain larger industries, and so supervision is done on a spot-check basis. Greater resources are needed. This would allow controls to be more efficient.

The waste which has the greatest chance of escaping control in large quantities is "better" waste oil. This has a high "market value" as fuel oil and is quite easy to sell, although this oil often contains dangerous impurities. The same applies to lubricating oil. Better and increased control has been possible now that all municipalities have introduced collection monopolies.

Uncontrolled hazardous waste can occur, however well laws and regulations are designed. Legislation can be evaded by less responsible generators. This type of uncontrolled waste occurs only in relatively small quantities, but can result in serious damage to the environment. Here too, increased control and better resources for the supervisory bodies are the antidote, but it is also important that the management of hazardous waste under the monopoly set up by law is performed in a service-minded manner and at the "right" cost for the market if the tendency towards uncontrolled management is to be minimized.

Future policies will be designed to reduce the generation of hazardous waste by changing industrial processes and increasing internal recycling and to streamline collection and control systems.

Apart from what has been mentioned concerning the efficiency of controls, illegal management by less scrupulous generators and capacity problems for certain types of waste, systems for the management of hazardous waste function relatively well in Sweden. The greatest deficiency is probably that the collection system does not yet extend to all parts of the country. This aspect must be developed, and high levels of competence must be maintained.

In spite of extensive legislation, it happens that contractors or companies with hazardous waste deliberately or inadvertently bypass the law. Disposal may be carried out in this way without the controls required by law. Certain problems with the definitions may also contribute to uncontrolled operations.

In such cases, greater resources must be provided for information and supervision. Otherwise, conscientious companies do not work under the same conditions as the less conscientious ones, and this leads to considerable discrepancies in the costs of hazardous waste management. This, in turn, could lead to reduced quantities of hazardous waste being dealt with by the legitimate organizations, with further negative effects for the cost of hazardous waste management and for the environment.

As has been mentioned, the cure for such undesirable development is — apart from improved supervision resources — the establishment of efficient and economically viable systems for the disposal of hazardous waste.

By establishing efficient systems which enable the flow of hazardous waste to be controlled actively and competently, from collection and transportation to reception, treatment and recycling, improved management of hazardous waste could be achieved at a lower total cost both for waste-generating industries and for society.

Another area which requires particular attention is the fear of the general public for — and resistance to — anything to do with hazardous waste. The best way to promote understanding in these matters probably consists not only of serious and — from an environmental point of view — sound consideration of the questions surrounding hazardous waste, but also of a more open and accurate supply of information, from both administrators and politicians. A higher level of public confidence in the handling of hazardous waste is necessary if the best possible planning and location of resources is to be attained.

13

Hazardous Waste Management in THE UNITED KINGDOM

DAVID C. WILSON

Environmental Resources Limited, London, UK

OVERVIEW

Legislation to control hazardous waste management in the UK was first introduced in 1972. Since that time, more comprehensive legislation and controls have been implemented. Extensive research has been carried out, and is still continuing, particularly into the behaviour of hazardous waste in the landfill environment. Many working groups representing all interested parties have been convened by the government and have issued guidance documents, codes of practice for the safe management of particular types of hazardous waste, and, most recently, a comprehensive code of good practice for landfilling waste. Also, a number of independent reviews of waste managment policy have been carried out, notably by a Select Committee of the House of Lords (the Upper House of the British Parliament) and by the Royal Commission on Environmental Pollution.

As a result of all these acitivities there has been significant progress in hazardous waste management during the last 15 years. The problems are now recognized, and we are beginning to have the nucleus of a scientific understanding of them. Operational practices have improved enormously and there are now some very professional waste management companies active in the field.

The central element in the UK system of control is site licensing, which applies to all treatment and disposal facilities of household, commercial and industrial waste. The license conditions applying to facilities handling hazardous waste will obviously be more stringent than for other waste, but the system of control is the same. Additional controls over hazardous waste apply primarily to transport, comprising a consignment note system to ensure that each load arrives at its designated destination. In addition, for such "special waste", it is also mandatory to keep records showing the location of deposits within a landfill site.

Hazardous waste facilities are primarily provided by the private sector, with licensing, inspection and enforcement being the responsibility of the public sector. This function is developed to local waste disposal authorities, which vary in size from county councils in England to smaller district authorities in Wales and Scotland. The priority given to waste disposal varies between authorities, so that there are wide variations both in the status of waste disposal officers in the management hierarchy and in the level of resource committment.

A small central Hazardous Wastes Inspectorate was set up in 1983 with the aim: "... to satisfy Environmental Ministers and, through them, Parliament and the public at large, that what is being done in respect of hazardous wastes management, especially disposal, is safe, environmentally appropriate, acceptable to reasonable public opinion, and economical and efficient from the standpoint of the waste producer."

The Inspectorate have found that existing controls are very unevenly applied across the country. In particular, the standard for site licensing, inspection and enforcement vary widely between waste disposal authorities. This variation means that a few landfill sites in particular are able to accept waste that is considered elsewhere unsuitable for landfill, thus resulting in serious distortions in the market place and a downward pressure on prices.

Despite these problems, and the need for continued research to increase further an understanding of the processes involved, the controlled co-disposal of hazardous waste with household or similar waste is considered to be technically a successful practice.

The principal aim of co-disposal is to take advantage of the capacity of household waste to attenuate the leaching of polluting constituents from the difficult waste and to aid degradation by biological processes. Careful control is necessary to ensure that these attenuation processes are not overwhelmed, and all difficult waste must be appraised for compatability with household refuse and other difficult waste before landfilling in this fashion. Extensive guidelines on co-disposal have recently been published as part of a wide-ranging review of good landfill practice.

Particular problem waste includes:

- PCBs, for which there is a shortfall of incineration capacity;

- asbestos, which is currently landfilled at many sites, but where concern is felt over the sterilization of the land and the possibility of future disturbance;
- acid tars, for which until recently there was no acceptable treatment or disposal option available; and
- drummed waste, for which it is now recommended that direct landfilling is unacceptable. The problem here is that effective enforcement is almost impossible.

To summarize, great progress has been made in improving standards of hazardous waste management during the last 15 years. However, to quote a Government Minister, further improvements can and must be achieved "through better planning, greater professionalism, stricter policing, and economic charging of waste producers."

Perhaps the main lesson for other countries is that no system of control can work effectively without proper enforcement.

NATIONAL CONTROL SYSTEM

History of hazardous waste management

Legislation to control hazardous waste management was discussed in various Government-sponsored working parties in the late 1960s and early 1970s. These deliberations were overtaken by events in early 1972, when a number of incidents involving the illicit disposal of drums containing solid cyanide residues on waste ground used by children hit the headlines in the national press. Under the pressure of public opinion, an emergency bill was introduced by the government, and passed through all its parliamentary stages in a record 10 days to become the Deposit of Poisonous Wastes Act.

This simple but effective piece of legislation required that producers should notify local waste disposal authorities in both their own area and in the destination area of their intention to treat or to dispose of any hazardous waste. The deposit of waste without such notification became a criminal offence.

The Control of Pollution Act became law in 1974. This is a wide-ranging enabling act, under which regulations have gradually been introduced over the last 10 years to cover various aspects of the control of waste management. For example, the licensing of all facilities handling household, commercial or industrial waste was introduced in 1976 and additional regulations controlling certain hazardous or "special" waste were enacted in 1980, enabling the Deposit of Poisonous Waste Act to be repealed.

Since 1974, the Department of the Environment has convened a number of working groups representing all interested parties in waste management, including producers, contractors, control authorities and professional bodies, to produce guidance on particular aspects of waste management and codes of practice for the treatment and disposal of particular types of hazardous waste. Some 15 of these codes of practice have now appeared.

Parallel to the development of legislation and Waste Management Papers, Central Government has invested considerable resources since 1972 in research programs, particularly directed at the behaviour of hazardous waste in landfill sites. A preliminary study by the Institute of Geological Sciences suggested that, of 2500 operational landfill sites in 1971, only 50 appeared to have the potential for causing ground or surface water pollution. Some 20 of these sites were investigated in detail in a major field research program from 1972 and 1976, which also included controlled lysimeter and laboratory experiments. The broad conclusion of this research was, "that sensible landfill is realistic and that an ultra-cautious approach to landfill of hazardous and other types of waste is unjustified."

Continuing research has been directed mainly at the practice of controlled co-disposal of certain hazardous waste in municipal waste landfills.

Immediately following the 1972 Act, considerable quantities of hazardous waste began to be concentrated on a few landfill sites. This led inevitably to allegations by local people that they were becoming a dumping ground for the nation's hazardous waste. In response to pressure from one particular local authority, and also to national publicity for complaints of health effects at a few landfill sites, the Select Committee on Science and Technology of the House of Lord instituted an inquiry into hazardous waste in 1980. They received evidence from all interested parties and their report is recommended reading. They broadly endorsed a policy of controlled co-disposal for many hazardous wastes, but made a number of far-reaching recommendations including a strengthening of enforcement and a thorough review of landfill practice.

One of the principal recommendations of the committee, that a small national inspectorate be set up to overview standards of operation of, and enforcement of controls over, waste disposal facilities was quickly accepted by the Government, and the Hazardous Waste Inspectorate (HWI) began operation in August 1983. The first report of the Inspectorate was published in 1985 (12) and the second in July 1986.

In response to another of the committee's recommendations, a Landfill Practices Review Group (LPRG) was set up in 1982, with 60 members including representatives of the public and private sectors of the landfill industry, consultants, the water industry, waste disposal authorities, research organizations, and several Government departments. To quote from the press previews of their report, "this breadth of expertise has allowed the consolidation of an internationally unique 200,000-word guidance document, which integrates the latest findings of landfill research with tried and proven operational practices". The report was published as Waste Management Paper No. 26 in August 1986.

The Joint Review Committee (JRC) is another Government sponsored consultative group set up to review the first year of operations of the Special Waste Regulations introduced in 1981 under the Control of Pollution Act. Its brief was expanded to consider some of the more far reaching recommendations of the House of Lords Committee, for example the registration of waste producers and licensing of contractors and transporters. Its report was published in May 1985.

The independent Royal Commission on Environmental Pollution (RCEP) is the lastest body to examine the state of waste management in the UK. Their report was entitled "Managing Waste: The Duty of Care" and was published in December 1985.

Several amendments to waste disposal law are now in the pipeline, largely in response to problems or loopholes highlighted by the various high level reviews discussed above.

Summary of legislation

The major provisions of the 1974 Control of Pollution Act as they affect hazardous wastes are summarized in Table 1.

TABLE 1
Provisions of the Control of Pollution Act 1974 as they Affect Hazardous Waste Management

Section No.	Date implemented	Provision
1	Not yet	Places a duty on waste disposal authorities to ensure that all arrangements for the disposal of controlled wastes are adequate.
2	1978	Places a duty on disposal authorities to conduct waste disposal surveys, to prepare waste disposal plans and to update them.
3–11	1977	Introduces a comprehensive system of site licensing to control all waste transfer treatment or disposal facilities.
12–13	Only in part	Defines the power of collection authorities, who have a duty to collect household waste and commercial waste if so requested.
14	Not yet	Governs co-operation between collection and disposal authorities.

TABLE 1 (continued)

Section No.	Date implemented	Provision
17	1981	Allows additional regulations to be introduced to cover "special" wastes. The present regulations are directed specifically at the transport of wastes which could be hazardous to human health if disposed of other than at a licensed site.
30	1974	Defines terms used in the act. "Controlled waste" means household, industrial and commercial waste or any such waste. "Special waste" is any controlled waste which is difficult or dangerous to dispose of. Controlled waste excludes wastes both from agriculture and from mines and quarries.

The key Section 1 of the Act, which would require Waste Disposal Authorities (WDAs) to ensure the availability of adequate disposal facilities in their areas, has not yet been implemented, largely due to continuing financial restraints on local government. Section 2, requiring each authority to carry out a survey and prepare a plan was implemented in 1977, but no timetable has ever been attached to it. In March 1984, only 23 out of the then 165 WDAs had published a draft plan for consultation, but better progress is now being made.

Sections 3 to 11 of the Act cover site licensing, and detailed regulations to implement this were introduced in 1976 and amended in 1977. Further amendments are due to 1987.

Licensing applies to all facilities handling controlled waste, including household, commerical and industrial waste. It is seen as providing generally adequate control for the treatment and disposal of hazardous waste, so that Section 17, which gives the Secretary of State discretionary powers to introduce additional controls over "special" waste, has been used mainly to introduce a trip ticket and manifest system controlling the transport of hazardous waste. The only additional control for disposal is to require the site operator to keep records of the location of special waste within the disposal site, which records should be retained by the WDA when the site license is subsequently surrendered or revoked.

Waste going for recycling has hitherto not been controlled waste, and are thus not subject to the site licensing and special waste regulations. However, recycling of mineral oils and solvents will be brought within the licensing system by the 1987 regulations.

Licensing of facilities

All treatment and disposal facilities for controlled waste require a license issued by the local Waste Disposal Authority. Guidance in the preparation of such licenses given in Waste Management Paper No. 4. License conditions can be very detailed, governing the types of waste acceptable, detailed operating procedures to be followed, monitoring to be undertaken, and other requirements.

The Hazardous Waste Inspectorate commissioned a survey of site licenses of all waste management facilities in England and Wales shortly after it was set up in August 1983. This survey revealed some 5256 valid licenses, although some of these applied either to facilities that had closed or had never started operations. The variation in the standards of site licensing between different waste disposal authorities is considerable, and the vagueness in many licenses do not allow reliable statistics to be derived from the survey on the number of facilities accepting hazardous or special waste. The HWI has undertaken a detailed study of site license conditions and their variation between WDAs, preliminary results of which were contained in their second report.

The original regulations exempted certain storage facilities from the licensing requirement. This exemption was seen to be a loophole following an incident where phenolic waste were imported from The Netherlands for temporary storage pending re-export, supposedly for use as a defoliant in developing countries. Following the prosecution and the liquidation of the exporting company in The Netherlands, the importing company also went into liquidation, leaving several local authorities around the coast of England with an embarrassing problem.

The 1987 amendments to the site licensing regulations brings storage of waste for longer than 7 days within the licensing requirements. Exemptions are limited largely to not more than 50 m^3 of solid waste for up to 28 days.

The consultation paper issued in reponse to recommendations by the Joint Review Committee and the Royal Commission also envisages provisions:

- to make clear that a breach of waste disposal site license conditions is an offence in its own right, unrelated to any deposition of waste;
- to empower WDAs to have regard to the qualifications and previous conduct of applicants when determining site license applications;
- to make clear that WDAs are entitled to impose site license conditions that relate to periods additional to those during which waste is being deposited at a site, i.e. post-closure care; and
- to enable WDAs to charge holders of site licenses with the administrative and enforcment costs incurred by the WDA.

Trip tickets and manifest system requirements

The regulations introduced under Section 17 of the Control of Pollution Act require all shipments of special wastes to be accompanied by a consignment note or trip ticket, designed to ensure that the waste arrives at its designated destination. The manifest is filled in by the waste producer or on his behalf by a contractor, and despatched by him to the Waste Disposal Authorities in both his own area and in whose area the destination facility is located. Each time the waste changes hands, that is from the producer to the transporter and from the transporter to the disposer, the manifest has to be signed and a copy retained. The operator of the receiving facility must send a copy of the manifest to the producer's Waste Disposal Authority.

The idea of the system is that the WDA in whose area the waste arises matches up forms received from the producer and the receiver so that if none materializes from the latter, action can be taken to trace the shipment. The form from the producer to the receiving WDA must arrive at least 3 clear working days in advance of the removal of the consignment. This is designed to enable that authority to monitor or, if it wishes, witness the disposal.

Roles of national, regional and local government

The responsibility for planning, site licensing, enforcement and administration of the consignment note system for transport lies with the Waste Disposal Authority.

By 1986, there were 165 Waste Disposal Authorities in the UK, comprising various levels of local government:

- The Greater London Council (GLC);
- The 6 Metropolitan County Councils in England;
- The 39 other County Councils in England;
- The 37 District Councils in Wales;
- The 53 District and 3 Islands Councils in Scotland; and
- The 26 District Councils in Northern Ireland.

Thus, WDAs comprised a mixture of large authorities in the metropolitan areas, county authorities in the rest of England, and smaller district authorities in Wales and Scotland.

With the control of hazardous waste treatment and disposal resting primarily on site licensing, the existence of so many waste disposal authorities, ranging from the small to the large, obviously gives scope for considerable variations across the country. It was largely for this reason that the Hazardous Waste Inspectorate (HWI) was set up to ensure uniformity of standards. Their first and second reports expressed concern at the wide variations in standards of site licensing, inspection and enforcement that they found.

As one would expect, the larger metropolitan authorities, and the Greater London Council, were among the more effective waste disposal authorities, particulary with respect to hazardous waste control. However, these authorities were abolished with effect from April 1986, so that responsibility for waste disposal has reverted to the (68) smaller metropolitan district and London borough councils. In practice, many voluntary or statutory groupings of authorities have been formed to coordinate activities. In particular, a single, centralized Hazardous Waste Unit has been maintained for each of the former metropolitan or GLC areas.

In the waste disposal industry, opinion from producers and contractors alike had supported the continuing existence of these large authorities. The House of Lords Select Committee had earlier recommended a general move towards regional groupings of waste disposal authorities, specifically to increase control over hazardous wastes, but the government has recently rejected this recommendation, preferring to limit regional co-ordination to informal consultation between neighboring waste disposal authorities.

The role of national government is seen primarily in the fields of:

- legislation;
- international activities;
- provision of administrative and technical advice to WDAs;
- adjudication of planning or site licensing appeals;
- commissioning of research and development; and
- via the HWI, ensuring uniformity of standards across the country.

Subsidies

The policy of central government in waste disposal is one of strict non-invervention in the market place. Almost all treatment and disposal facilities for hazardous waste are provided by the private sector and receive no direct or indirect subsidy.

Most landfill sites accepting hazardous waste are run by the private sector, although some waste disposal authority sites do receive hazardous waste for co-disposal. In some such cases, the question of an indirect subsidy may arise.

If standards of site licensing were uniform, so that distinct categories of waste had to be disposed of by incineration or treatment, then this market system could in principle work well. However, standards vary widely, and to quote a ministerial statement, "These effects are introducing serious distortions in the market place, resulting in declining prices. This is leading to an undesirable trend away from treatment towards direct landfill and a general levelling down in standards".

Responsibilities of waste producers, transporters and facility operators

Table 2 summarizes the existing responsibilities of the parties involved in hazardous waste management.

TABLE 2
Responsibilities of Waste Producers, Transporters and Facility Operators with Regard to Special Waste

Waste producer

(1) Decide if his waste comes within the control system

(2) Prepare a manifest (consignment note) to accompany each load of waste and return advance and dispatch copies to WDA(s)

(3) Keep a register of consignment notes

(4) Obtain a site license for any treatment or disposal on his own premises

Waste transporter

(1) Deliver the waste only to a disposal sites licensed to receive it

(2) Transmit to the disposer the appropriate copies of the consignment note

(3) Keep a register with copies of consignment note relating to waste removed

(4) Handle waste so as to ensure the absence of risks to health and safety (Health and Safety at Work Act 1974)

(5) Comply with the regulations in force covering road tanker and tank container vehicles, e.g. marking and accompanying documents

Facility operator

(1) Have a valid site licence to accept the particular waste and operate in compliance with it

(2) Complete Part E of the consignment note, retain one copy, give one copy to the carrier and send one to the authorities of the area where the waste was generated

(3) Keep a register of consignment notes

(4) For every deposit on a landfill site, keep a special register of actual disposal locations and on termination of business activity forward such a register to the waste disposal authority.

In response to recommendations by the House of Lords Select Committee on Science and Technology, the Joint Review Committee and the Royal Commission on Environmental Pollution, the recent consultation paper is proposing to extend the Control of Pollution Act:

- to enhance the responsibilities of waste producers so that they have a general obligation to take all reasonable steps open to them to secure satisfactory disposal, the so-called "duty or care"; and

- to require all handlers of waste to register with the waste disposal authorities in whose areas they operate.

Programs for dealing with old landfill sites

There is little evidence for the inheritance of problem landfill sites in the UK on the scale reported in some other countries. Those that do exist have been dealt with as problems have arisen. Particular problems have been caused by the landfilling of acid tars.

A number of reasons may be suggested for this relatively favorable situation. Land use planning controls were first introduced at the end of the nineteenth century, and comprehensive controls have existed since 1949, so that indiscriminate use of land for waste disposal has not been possible.

As mentioned, specific legislation controlling hazardous wastes was first introduced in 1972, and site licensing has been in force since 1976.

Another positive factor is that the geology of the country is relatively favorable. Thus, only 50 out of 2500 operational sites in 1971 were judged as posing a potential threat of water pollution. Finally, most of the country is supplied with piped mains water, and the responsibility for the protection of groundwater is vested in statutory bodies.

HAZARDOUS WASTE DEFINITION

In the UK, there is currently no legislative definition of the term hazardous waste, because the control system does not require such a definition.

Thus, controlled waste, covering all household, commercial and industrial waste is subject to the general controls of the Control of Pollution Act, including the requirements for site licensing. Additional control of transport is reserved for those wastes that, if illegally disposed of between their point of arising and their proper place of disposal, could prove dangerous to life. The definition of a special waste is thus framed specifically to isolate only those hazardous wastes potentially dangerous to life if encountered after fly-tipping. According to the special waste regulations, "Waste is to be regarded as dangerous to life for the purposes of these regulations if a single dose of not more than 5 cm^3 would be likely to cause death or serious damage to tissue if ingested by a child of 20kg body weight, or if exposure to it for 15 minutes or less would be likely to cause serious damage to human tissue by inhalation, skin contact or eye contact."

The defintion also includes wastes with a flash point of 21°C or less, and medicinal products.

The original Deposit of Poisonous Wastes Act passed in 1972 contained a very simple definition of a hazardous or "notifiable" waste. Thus, all wastes were notifiable, except those on a specific list of exclusions, which covered a wide range of non-hazardous and biodegradable waste. Since the requirement of the act concerned mainly the notification of the intention to deposit such waste, the penalty for including some relatively harmless waste in the definition was minor.

The major types of waste, covered by the now defunct definition of a notifiable waste and which are not special waste, are those where the primary risk is one of water pollution. An example is heavy metal containing waste. Environmental protection following the disposal of such waste should be ensured by appropriate conditions on the site license.

SOURCES, QUANTITIES AND TYPES OF HAZARDOUS WASTE

Data sources and global estimates

There is no statutory requirement for producers to notify quantities of hazardous waste to the Waste Disposal Authorities. Proposals for registration of waste producers and mandatory reporting have been made by local authorities and endorsed by the House of Lords Select Committee, but these were rejected by industry and government representatives on the Joint Review Committee.

Under the Control of Pollution Act, authorities are obliged to carry out a survey of waste arising in their area, but due to financial constraints progress in this direction has been slow and variable across the country. Survey information is currently available from about half of the Waste Disposal Authorities in England, but the dates of these surveys range from 1977 to 1986 and the classification and reporting schemes used vary widely. Compiling definitive statistics is thus difficult.

The House of Lords Select Committee circulated a questionnaire to all Waste Disposal Authorities in 1980, and their report included a compilation of statistics of hazardous waste in 1979. However, these data need to be interpreted with care as the lack of a clear definition of hazardous waste resulted in considerable anomalies between different authorities. For example, one country included in its total, effluent discharged to sewer.

The Hazardous Waste Inspectorate has carried out a similar exercise annually since 1985, obtaining information from most Waste Disposal Authorities in England and Wales on total quantities of special waste. Their estimates totalled 1,650,000 tonnes in 1984 and 1,500,000 in 1985. If a figure of total hazardous waste of 2.5 times special waste is estimated (as suggested by the Joint Review Committee on operation of the special waste regulations), then the total quantity of hazardous waste is estimated at 4.4 million tonnes (1984) or 3.7 million tonnes (1985).

For total hazardous waste, the distribution between various treatment and disposal routes is estimated as follows (see Figure 1):

	1984	1985
Landfill	85.1%	78.7%
Solidification	3.2%	3.6%
Mine shafts	2.3%	1.6%
Sea disposal	4.7%	7.0%
Chemical treatment	3.4%	7.5%
Incineration	1.3%	1.6%

These statisics do not include figures for recycled materials, which until the proposed 1987 regulations have not been included within the definition of controlled wastes. It is estimated that 200,000 tonnes per year of solvents are recycled by the commercial sector, so that one could restate the total quantity of special waste as nearly 2 million tonnes per year of which some 15% is recycled. In addition, some 170,000 tonnes per year of waste oil are recycled, including use of waste oil as a fuel.

Breakdown of overall waste quantities

A further breakdown of these estimates by producing industry, waste type or physical form is very difficult. Some estimates for particular types of waste are contained in the individual Waste Management Papers. A compilation of such information was contained in a Department of the Environment Report to the Commission of the European Community some years ago, but all these data are now very out of date.

In principle, detailed information on industrial and hazardous waste should be available from the surveys carried out by waste disposal authorities under the Control of Pollution Act. However, as noted above, not all authorities have carried out surveys and the quality of those available is very variable.

Environmental Resources Limited carried out an ad hoc comparison of waste disposal survey results in mid-1985. Published surveys or plans were

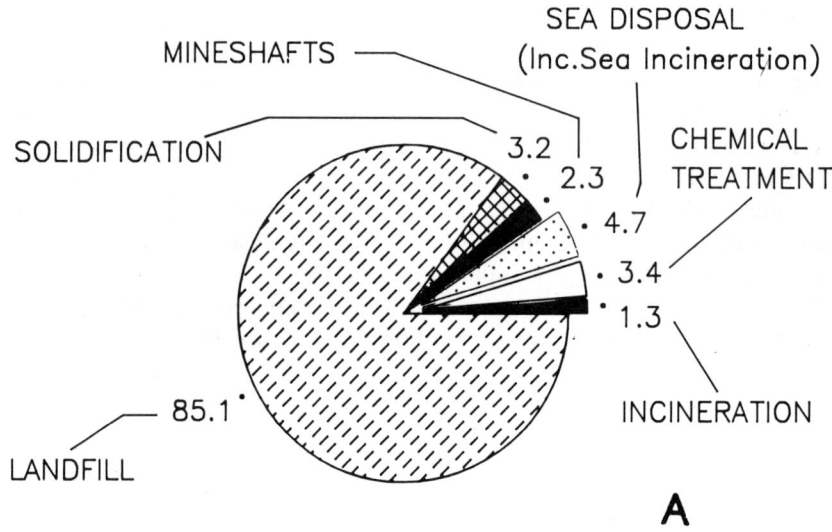

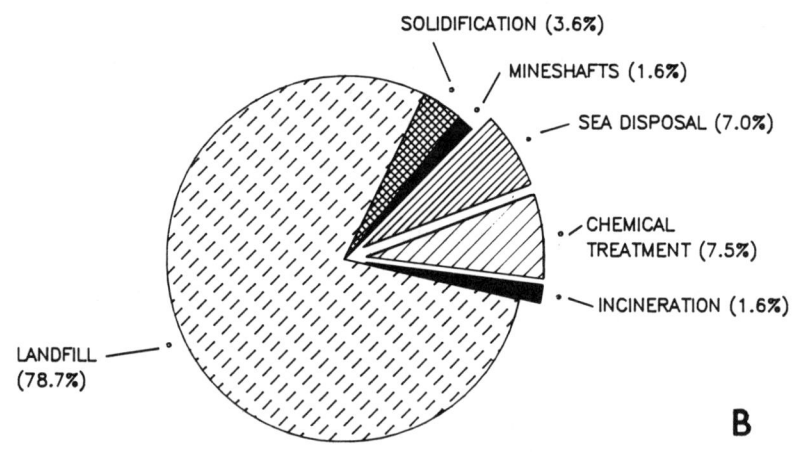

FIG. 1 Hazardous waste: disposal routes. Percentages of hazardous waste to each route in England and Wales.
A: 1984 — percentages of a total of 4.4 million tonnes.
B: 1985 — percentages of a total of 3.7 million tonnes.

Crown Copyright. Reproduced with the permission of the Controller of HMSO.

obtained from 19 waste disposal authorities in England, representing the majority produced at that time.

The level of detail on waste types and quantities given in these reports may be illustrated as follows:

- 16 (out of 19) give a breakdown of industrial waste by industry-type (SIC category);
- six give similar information for hazardous waste;
- 12 give some breakdown of hazardous waste by waste type;
- nine give breakdown of industrial waste by physical form (solid, liquid or sludge); and
- six give information for hazardous waste.

This wide variation in both the level and the quality of reported information, together with the ad hoc nature of the sample of counties examined, which may not be representative of the mix of waste types or industry types across the country as a whole, make the utilization of the data almost impossible.

To give one example of the sorts of difficulties encountered, the percentage of total industrial waste arising from the chemical industry varies between counties from 1–30%, with one county reporting a figure of 95%. The absolute quantities reported by this county are in excess of 20 million tonnes per year, which is of the same order as the aggregate from all the other counties.

When the figures are examined closely, it appears that perhaps 90% of the chemical industry waste from this county comprises water, there being extensive use of surface lagoons as settling ponds prior to discharge of effluents to water courses. In addition, the estimation method used in the survey appears to have been based on maximum quantities permitted under site license conditions; a number of sites are only used on an emergency basis, being operational for perhaps 20 days per year, but for which the licensed quantity may be as much as 2000 tonnes per day. The total estimate is therefore not comparable with information from other authorities.

Such information as is available on the distribution of hazardous waste by physical form is presented in Table 3. The only meaningful comment on this table is that liquids and sludges tend to predominate over solids.

Regarding information on the breakdown of hazardous waste arising by waste type, comparison is made difficult by the different classification systems used by the different authorities. The best comparison we could obtain was that using the Deparment of the Environment waste categories. Information for the nine authorities reporting using this system is shown in

TABLE 3
Hazardous Waste — Physical Form

Waste Disposal Authority	Quantities			Percentage		
	Solid	Liquid	Sludge	Solid	Liquid	Sludge
Cleveland	16,166	294,157	83,790	4	75	21
West Midlands	112,240	307,440	87,840	23	63	18
Tyne & Wear	23,639	40,369	3457	35	60	5
Cheshire	96,581	358,409	320,105	12	46	42
Northumberland	9780	1160	260	81	10	9
Dorset	360	1940	(see solids)	17	83	(see solids)
Suffolk	433	12,790	9644	2	56	42
Norfolk	1182	27,747	3409	4	86	10
Total	260,381	1,044,012	508,505	14	58	28

Table 4. This table is included simply to show the generally poor level of data availability. No attempt has been made to draw any generalizations as to the average distribution of hazardous wastes by type.

COLLECTION AND TRANSPORTATION SYSTEMS

Introduction

Collection and transportation of hazardous waste in the UK is carried out mainly by the private sector. Licensing of all contractors (and subcontractors) involved in collection and transport was recommended by the House of Lords Select Committee in 1981, but the Joint Review Committee, reporting in 1985, was unable to agree on a common position. The government has proposed a simple registration scheme. The information to be supplied includes: details of all convictions within the last 5 years for relevant offences; a statement of the classes of waste the registrant is prepared to handle; and details of insurance cover.

Statutory controls over collection and transportation of hazardous waste include:

- labeling of vehicles in accordance with general regulations on conveyance of hazardous substances;
- a transport emergency card giving information in the event of an accident should be carried in the vehicle;
- for special waste, a consignment note identifying the waste should travel in the vehicle; and
- it is the responsibility of the contractor to deliver the waste only to a facility licensed to receive it.

TABLE 4
Hazardous Waste Arising by Type (tonnes per year)

Department of the Environment Waste Category	Somerset	West Midlands	Cheshire	Cleveland	Tyne & Wear	Cambridgeshire	Dorset	Staffordshire	Northumberland
A Inorganic acids	398	154,670	6200	288,020	1484	200	830	486	
B Organic acids and related compounds			3320		928	150		630	350
C Alkalis	<100	27,730	14,901	190	25	40	10	1638	
D Toxic metal compounds	77	23,600	264,631	1220		66	20	1326	30
E Non-toxic metal compounds		5620	8374	1223				198	
F Metals (elemental)		90							
G Metal oxides			367						
H Inorganic compounds	30	7830	5470	44,700	10		20	144	9610
J Other inorganic materials	280	75,580	338,790		60,600		120	210	50
								7260	
K Organic compounds	1865	7000	3369	7000	3800	1780	390	750	630
L Polymeric materials and precursors		7810	2591	13,200	140	220	<10	324	
M Fuels, oils and greases	470	96,530	8501	20,200	3700	3200	<10	2220	530
N Fine chemicals and biocides			42,800		160	1400	280		
P Miscellaneous chemical waste	30	390	3375			130	40		
Q Filter materials, treatment sludge and contaminated rubbish		41,110	62,267				340	10,020	
R Interceptor wastes, tars, paint, dyes and pigments	123	21,810	12,291		2500	80	250	1728	
S Miscellaneous wastes	1760	18,550	395	11,000					
T Animal and food wastes		70	69			*			

* 8541 tonnes per year of "special hazard" waste not entered

Consignment note system

The form of a statutory consignment note is shown in Figure 2. The procedure for completing the forms is outlined in the official guidance document on interpretation of the regulations as follows:

(a) "the firm producing the waste will complete on the first form in the multipart set those sections (Part A and B) giving details of the waste, the place and process from which it orginated and the site or plant to which it is to be sent. The top copy should be detached and sent to the Waste Disposal Authority for the area in which the intended (and suitably licensed) disposal site is situated. It must arrive not more than 1 month and at least 3 clear working days in advance of the removal of the consignment. This will enable the authority to monitor or if it wishes, witness the disposal.

(b) the carrier will complete Part C on the second form which records his receipt of the waste. There is space here for recording any discrepancies between the carrier's perception of the consignment and the producer's description. (For instance, the producer may have described the consignment in tonnes where the carrier may only be able to verify the number of containers). The producer will then complete Part D on the second form which records the carrier's collection of the waste and will verify any amendments made by the carrier to the description of the waste. If the disposal site is in a different area from the one in which the waste producer is located, the second form should be send by first class post, at the time of despatch to the Waste Disposal Authority for the area in which the waste was produced.

(c) the producer will detach the third form for his records and hand the remaining copies of the partially completed form to the carrier.

(d) at the end of his journey, the carrier will hand the remaining copies of the form in his possession to the disposal firm, which should then record receipt of the waste on the fourth form at Part E. The carrier will retain the fourth form; the fifth form should go to the Waste Disposal Authority for the area in which the waste arose by first class post not later than one working day after the disposal of and waste and the remaining form should be retained by the disposer for his records."

It is the responsibility of the Waste Disposal Authority for the area in which the waste is produced to ensure that all these stages are completed. If the authority is not so satisfied, it will use the information on its first copy of the form to trace the load, and the responsibility for it, from its last known location.

Special collection systems

There are no general requirements in the UK requiring waste disposal authorities to make arrangements for the collection of small quantities of special waste. However, some arrangements do exist in some areas. As to

Department of the Environment/Welsh Office/ Scottish Development Department	Serial No.
CONSIGNMENT NOTE FOR THE CARRIAGE & DISPOSAL OF HAZARDOUS WASTES	

Producer's Certificate A	(1) The material described in B is to be collected from................................ and (2) taken to.. Signed.......................... Name.......................... On behalf of.................... Position....................... Address and telephone......... Date........................... .... Estimated date of collection..........

Description of the Waste B	(1) General description and physical nature of waste (2) Relevant chemical and biological components and maximum concentrations (3) Quantity of waste and size, type and number of containers (4) Process(es) from which waste originated

Carrier's Collection Certificate C	I certify that I collected the consignment of waste and that the information given in A(1) & (2) and B(1) & (3) is correct, subject to any amendment listed in this space: I collected this consignment on............ at......:...... hours Signed.............. Name.............. Vehicle Registration No...... On behalf of.................................. Address and telephone.................. Date............
Producer's Collection Certificate D	I certify that the information given in B & C is correct and that the carrier was advised of appropriate precautionary measures. Signed.............. Name.............. Telephone.............. Date..............
Disposer's Certificate E	I certify that Waste Disposal Licence No.............. issued by .... Council, authorises the treatment/disposal at this facility of the waste described in B (and as amended where necessary at C). Name and address of facility This waste was delivered in vehicle (Reg. No.).............. at......:...... hours on (date).............. and the carrier gave his name as.............. on behalf of.............. Proper instructions were given that the waste should be taken to.............. Name.............. Position.............. Signed.............. on behalf of.............. Date..............
For use by Producer/ Carrier/ Disposer	

FIG. 2 Statutory form for a special waste consignment note. Crown Copyright. Reproduced with the permission of the Controller of HMSO.

waste oil, all waste disposal authorities provide "civic amenity" sites, where householders may deliver their wastes. At many of these sites, special containers are provided for waste oil. In addition, many garages provide waste oil reception facilities as a service to their customers. There are no statutory requirements that such facilities must be provided, and their provision essentially relies on market forces.

Concern has been expressed by the HWI at the high level of chlorinated solvents in some samples of waste oil destined for use as heating oil. One survey showed contamination in the range of 0.1 to 0.4%, sufficient to present a cause for concern over pollution from the boiler.

Arrangements for the collection and reception of hazardous household waste are even more patchy than for waste oil. Some civic amenity sites provide special containers for the receipt of asbestos-containing waste, and some local authorites will collect such waste from households or construction sites. From time to time, campaigns are organized through pharmacists for the collection of surplus medicines from households. The Department of Health and Social Security supply most mercury-containing hearing aid batteries, and new batteries are only supplied in exchange for old. The returned batteries are collected at a central point and sold to a specialist recovery firm. However, it is understood that reclamation now takes place, not in the UK, but in West Germany.

The disposal of small quantities of a wide variety of chemicals from laboratories in schools and other educational establishments is a considerable problem. A few local authorities, particularly those in metropolitan areas, offer a collection and disposal services. A national service is operated on a commercial basis by the Atomic Energy Authority, but the economic price for handling small quantities of sometimes very difficult materials is high.

Marine waste

The Marpol Convention of 1973 governs marine pollution by ships on an international scale. Annex 1 of the Convention was introduced in October 1984, and requires that washwater from oil tankers should be discharged to tanks in port for subsequent treatment.

In the UK, most of the oil is handled by large companies who have provided adequate facilities for some time. Major new facilities for the treatment of washwater from North Sea Oil facilities were built around 1980. Under the Convention, if a ship complains that facilities at a particular port are inadequate, then the authorities must investigate the complaint and ensure that facilities are brought up to standard. So far, no such complaints have been received.

Annex 2 of the Convention covers washwaters from certain chemical tankers and comes into force in April, 1987.

13. THE UNITED KINGDOM

TREATMENT, STORAGE AND DISPOSAL SYSTEMS

Introduction

Table 5 shows an approximate categorization of licensed facilities in England and Wales into some 20 types of facilities. The distinction between sites receiving some hazardous wastes and those receiving only non-hazardous wastes is approximate, as many site licenses use only vague and general descriptions to indicate the types of waste which may be accepted.

A number of more particular comments may be made from the data in this table. For example, the vast majority of the facilities listed as handling hazardous waste receive only small quantities, often on an occasional basis. Also, some of the facilities are licensed for only one type of hazardous waste, commonly asbestos. This applies to most of the transfer stations and civic amenity sites listed, as well as to a significant proportion of the landfill sites.

TABLE 5
Licensed Facilities in England and Wales (extracted from HWI's Register of Licensed Facilities. Nos to March, 1984)

Facility type	Hazardous	Non-hazardous	Total
Landfill	1145	3057	4202
Transfer station	138	211	349
Storage	99	4	103
Incineration	92	75	167
Lagoon	53	37	90
Recovery	47	3	50
Treatment	39	5	44
Civic amenity	22	136	158
Mineshaft	14	6	20
Sludge land farm	9	5	14
Reception pit	5	0	5
Baling	4	7	11
Soakaway	3	0	3
Solidification	2	1	3
Mech. dewaterer	2	0	2
Waste derived fuel	2	1	3
Aerosol destructor	1	0	1
Pulverization	1	18	19
Evaporation pit	1	0	1
Composting	0	2	2
Unknown	3	6	9
Totals	1682	3574	5256

Crown Copyright. Reproduced with the permission of the Controlled of HMSO.

Similarly, some of the incinerators listed as hazardous are municipal facilities receiving only small quantities of, say, laboratory wastes from schools, for co-combustion with domestic refuse. Finally, the list includes both public sector (362) and private sector (1320) facilities. The latter comprise both contractor operated, and thus publicly available, facilities, and in-house facilities reserved for use by the operator.

The remainder of this section examines the available technologies in turn. The discussion is largely descriptive. Technical standards for the design and operation of facilities are not mandatory, being included in the site license, which is specific to each facility.

The number of publicly available facilities for hazardous waste is better indicated by Table 6. The number of "major" landfill sites is an estimate of those sites handling more than 5000 tonnes per year of hazardous wastes. The other figures are believed to be accurate as of early 1986. Of these facilities, all except a very few landfill sites are provided by the private sector. A map showing the location of specialist facilities is given in Figure 3.

TABLE 6
Publicly Available Facilities for Hazardous Waste Disposal

Facility Type	Number of Facilities
Major landfill sites (handling more than 5000 tonnes per annum)	35
Incineration and chemical treatment	2
Other incineration	3
Chemical treatment	11
Solidification	2
Mine shaft	1
Sea disposal terminals	10

The remainder of this section examines the available technologies in turn. The discussion is largely descriptive. Technical standards for the design and operation of facilities are not mandatory, being included in the site license, which is specific to each facility.

Storage

Table 5 lists 99 hazardous waste storage facilities as licensed in England and Wales. This may understate the true total, as prior to the introduction of the 1987 regulations facilities storing waste temporarily for periods up to 28 days were exempted from licensing.

The majority of these storage facilities are believed to be in-house. Little further information is available.

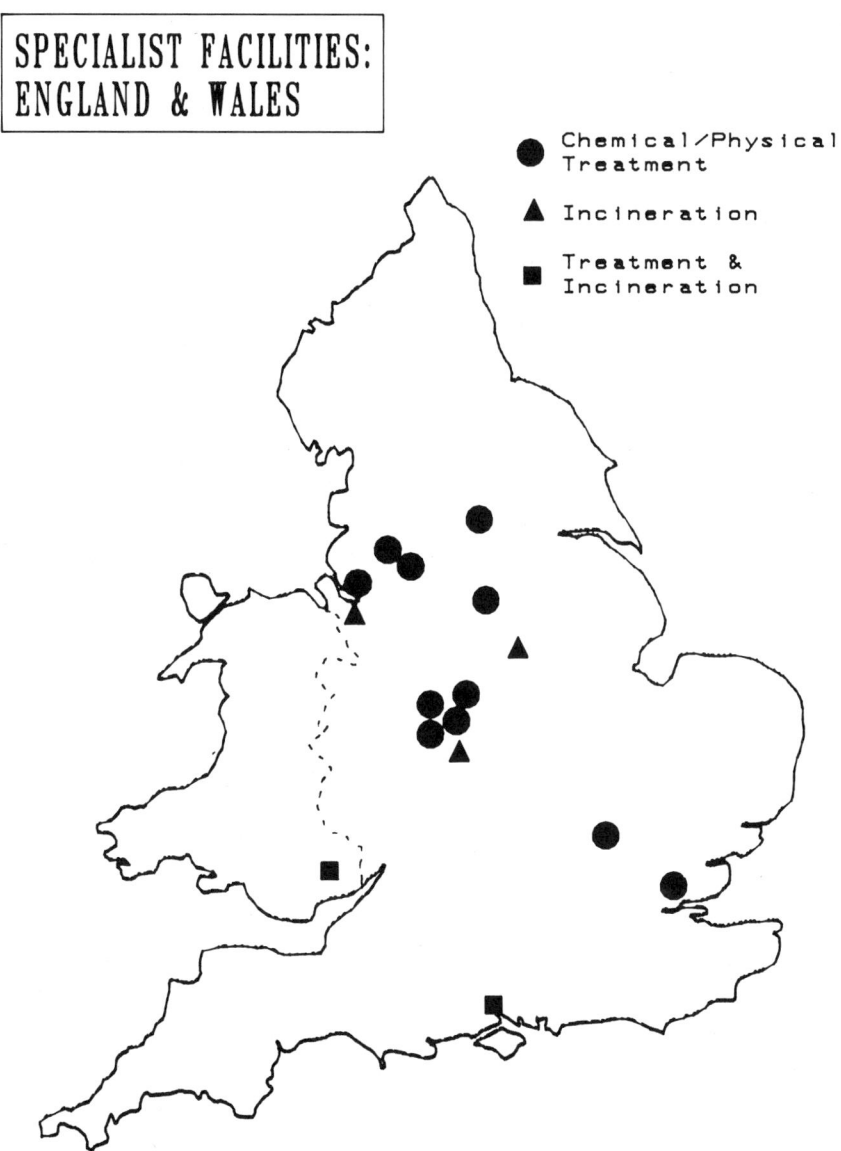

FIG. 3 Specialist waste treatment facilities in 1985. Crown Copyright. Reproduced with the permission of the Controller of HMSO.

Incineration

Available capacity

All chemical waste incinerators must be registered with the Industrial Air Pollution Inspectorate. Some 60 in-house and four commercially available or merchant incinerators are so registered. Also, one in-house plant accepts 4000 tonnes per year of liquid waste on a commercial basis. In addition, about 25 essentially non-hazardous incinerators are registered to burn small quantities of specific hazardous waste.

All the merchant facilities are unsubsidised private sector operations. Competition, at least in part from less satisfactory and cheaper disposal routes, has already led to seven merchant incinerators closing since 1974.

All four remaining facilities are equipped with wet scrubbing facilities, while two are licensed to burn PCBs. Only one of the four can accept significant quantities of solid waste. The total design capacity of the four plants plus the one in-house facility mentioned above is 65,000 tonnes per year including only 6000 tonnes for solid waste. This leaves little margin to deal with any increase in the 1985 UK market demand of 56,000 tonnes per year.

The HWI reported in 1985 that there was a shortfall in capacity for solid hazardous waste, particularly capacitors and transformer carcasses containing PCBs for which the waiting time was many months. This backlog had cleared by 1986. However, one consequence is that incineration of contaminated soil, practised in some countries, could not be considered in the UK due to limitations on capacity.

Types of incinerator

To my knowledge, there are no rotary kiln incinerators for hazardous wastes operation in the UK.

All four of the merchant incinerators employ different designs:

- a multi-chamber incinerator burning both liquids and sludges, which are fed continuously, and solids, which are loaded through a fire door. The temperature is maintained between 1000 and 1100°C, and the minimum residence time for materials injected into the first cell is 3 seconds. This facility burns PCBs in addition to other wastes;

- a modified liquid-injection incinerator, incorporating a hearth where some solids may be introduced;

- a modified vortex-type liquid waste incinerator with pumping and burning equipment specially designed to handle sludges; and

- another design of liquid waste incinerator, which is also licensed for PCB combustion.

The incineration of PCBs at two sites, one of which closed in September 1984 for commercial reasons, has been the subject of intense local opposition, concerned with alleged emissions of dioxins and PCBs. High level independent inquiries have shown no abnormal health effects, and research has demonstrated PCB destruction efficiencies better than 99.999%, with PCDDs and PCDFs below analytical detection limits (currently 1 part in 10^{12} in stack gases). The Hazardous Wastes Inspectorate thus regard high-temperature incineration as still the best practicable environmental option (BPEO) for PCBs and many other hazardous wastes, and it continues to sponsor research in this area.

Limited quantities of waste are incinerated in the North Sea, utilizing commercially available incinerator ships controlled under international convention. The quantities are relatively small, but increasing: 1981-810 tonnes; 1984-1950 tonnes; and 1985-2584 tonnes. These actual quantities incinerated are less than the licensed quantities. Nearly 8000 tonnes was licensed for incineration at sea in 1986.

Chemical treatment

Chemical treatment is commercially available at a total of 13 facilities. In 1984, installed treatment capacity (300,000 tonnes per year) greatly exceeded current market demand (150,000 tonnes per year). In addition, some 25 in-house facilities are licensed.

The majority of these sites operate a single process or a limited range of processes. A small number of the commercially available sites operate a range of processes:

- oil-water separation and cracking of emulsions, using a number of alternative processes;
- neutralization of acids and alkalis;
- thickening and de-watering of sludges;
- precipitation, for example the heavy metals;
- reduction of chromium-VI to chromium-III; and
- oxidation of cyanide.

There are two facilities offering solidification, primarily of inorganic materials, using a cement-based process.

Recent developments include the introduction of three new treatment processes, each of which may have wide applicability.

A new plant for treating acid tar wastes, arising largely from oil re-refining operations but also from white oil production and from benzole washing, has recently been comissioned. The plant capacity is 10,000 tonnes per year and the acid tars — a mixture of viscous organic material and concentrated sulfuric acid — are converted to an inert solid tar and an aqueous organic effluent that is discharged into a redundant mine shaft on the same site. A scrubber has been fitted to control sulfur oxide emissions from the process.

A new process known as "Vitrifix" converts fibrous asbestos into harmless glass in a specially adapted high temperature furnace. The volume of the waste is reduced by a factor of up to 10, and the product can be used as hardcore.

This process won its developers a Pollution Abatement Technologly Award in 1984, but its future has been in question as the government is unwilling to subsidise a full-scale demonstration of the process. However, it has been used to treat asbestos wastes at a former dump site being redeveloped in Scotland, for which off-site disposal was politically difficult.

A full-scale commercial plant is currently nearing completion for the pyrolysis of waste tyres, and their conversion to fuel products.

Landfill

Landfill of hazardous waste is still the predominant method of disposal, accounting for 85% of an estimated total of 4.4 million tonnes in 1984 and 79% of 3.7 million tonnes in 1985.

As pointed out earlier, much research has been undertaken during the last 15 years on the behavior of hazardous waste in landfill sites. Comprehensive guidance on good landfill practice has been published recently as Waste Management Paper No. 26.

There are broadly three types of landfills that can be adopted for hazardous waste:

- mono-diposal is suitable for large arisings of particular kinds of waste, for example from the bulk inorganic chemical or metal smelting industries;

- co-disposal of hazardous with non-hazardous bulk waste, commonly domestic refuse; and

- multi-disposal, where many different types of chemical wastes are disposed of together.

In many countries, the third option of multi-disposal, in a special chemical waste landfill site, is preferred. In the UK, this type of operation is viewed as generally undesirable, owing to the difficulty of ensuring that incompatible wastes do not react together, and to the need for long-term sterilization of the site. The option effectively amounts to long-term storage, and action is required to protect against future disturbance by redevelopment.

The principle of co-disposal is to mix carefully controlled quantities of particular hazardous waste with domestic or similar waste that has a capacity to attenuate significantly the leaching of the polluting constituents and/or to aid their degradation by biological processes.

Co-disposal is not suitable for all waste. Much further research is required to "firm up" on guidance regarding loading rates, but the Landfill Practices Review Group (LPRG) report (WMP No. 26) is a large step in the right direction.

The basic principle underlying many of the loading rates suggested by the LPRG is that co-disposal of hazardous with municipal refuse should not increase the concentration of hazardous constituents in the leachate above those already present in leachate from purely domestic landfills. The rates suggested as a guide are generally conservative, in several cases representing an average concentration over the landfill site of a few grams of active constituent per tonne of waste. In all cases, monitoring is essential to allow loading rates to be tailored to local conditions during the life of the site.

It should also be pointed out that many of the wastes advocated for co-disposal in the UK are commonly landfilled in other countries. A prime example is asbestos, which is not defined as a hazardous waste in the USA, and over which there is therefore no specific control related to disposal. About 50% of the co-disposal sites in the UK are licensed only for co-disposal of asbestos waste.

Another example is co-disposal of metal hydroxide sludges from the treatment of effluents from the metal-finishing industry. At moderate loading rates, such materials are effectively retained as insoluble compounds under the anaerobic conditions within a domestic landfill site. Site licenses generally prohibit the deposit of such sludges within several metres of the finished surface of the landfill, to prevent any effect on future agricultural use.

Co-disposal is a technically sound practice, but its successful implementaton requires good management and proper control. There is ample evidence from the first two HWI reports that this is often not the case at present.

Other land or subsurface disposal

Other land disposal options are relatively unimportant in the UK, despite the significant number of facilities listed in Table 5.

Lagoons are generally in-house facilities used as intermediate treatment plants, generally as settlement basins. Typically the supernatant aqueous layer is discharged to sewers, and the sludge either allowed to accumulate in the site or else allowed to dry prior to intermittent removal for disposal elsewhere. There are, as with all other facilities, no specific regulations regarding the lining of lagoons, each specific site being treated on its merits.

Deep well injection of waste is not utilized in the UK.

Land farming of refinery sludges takes place to a limited extent.

Fourteen *mine shafts* are listed in Table 5. These are generally in-house facilities utilized for particular types of waste. They include the re-injection of saturated brine into salt cavities, which are also utilized for the disposal of brine mud from the manufacture of soda ash and for intractable residues from chlorinated organics production. There are five contractually available facilities, perhaps the most significant of which accepts aqueous wastes contaminated with organic materials, the geology of the site being such that total containment is claimed. This facility re-opened in 1984 following a public inquiry.

Sea disposal

Some 260,000 tonnes of hazardous waste were dumped at sea in 1985, an increase from 205,000 tonnes in 1984. There are currently 10 licensed terminals and some 80 individual licenses for specific waste streams, not all of which are hazardous. Most of these wastes are essentially aqueous effluents, containing suspended solids and up to a few percent of generally biodegradable organic compounds such as phenols.

Recycling

The major hazardous wastes recycled "beyond the factory fence" are as follows:

- about 130,000 tonnes per year of solvents;
- 170,000 tonnes per year of waste oils;
- some heavy metals from the metal finishing industry;
- some mercury wastes; and
- silver from photo-finishing operations.

The Chemical Recovery Association (CRA) represents some 50 member companies in the solvent and waste oil recycling industry. The figures above are quoted from their statistics, and so do not include smaller operators. The HWI estimate the total quantity of solvent recovery at 200,000 tonnes per year, some of the feedstock being imported.

Even this figure grossly underestimates total solvent recycling, as very large volumes are recylced in-house, for example in the pharmaceutical industry where individual companies may each recycle in excess of 100,000 tonnes per year. Similarly, the figure for waste oil does not include in-house use as a fuel.

The solvent recovery industry is fiercely competitive, with many small operators who do not belong to the CRA. The HWI are concerned that these operators are treating the less tractable feedstocks, containing 30% or more of solids, and that their economics do not allow for the proper disposal of their residues. It is for such reasons that licensing is to be extended to oil and solvent recycling operations.

The current extent of mercury recycling in the UK is uncertain. A number of companies have ceased such operations, and mercury batteries are now exported to West Germany for processing.

Waste exchange schemes have not "taken off" in the UK. A passive national scheme closed after 5 years as it could not become commercially self-supporting. An active waste exchange was operated by the West Midland County Council, based on knowledge gained in the course of its pollution control activities.

Waste reduction

There is no nationally sponsored programme aimed specifically at reducing hazardous waste generation. The Department of Energy fund a series of demonstration projects in the field of energy recovery from waste, but few if any of these involve hazardous waste. The European Community have a similar cleaner technology demonstration programme, which obviously overlaps the question of hazardous waste reduction at source.

Import/Export

Total imports of waste into the UK in 1985 were estimated at 25,000 tonnes, most of it for incineration and chemical treatment. There is, however, evidence of a growing trend to import wastes for landfill, perhaps after minimal treatment. Substantial quantities of PCBs were imported for incineration. The major exporting countries were Ireland and The Netherlands.

Exports from the UK are very limited, with a small quantity, approximately 1000 tonnes per year, going to the salt mine in West Germany.

Prices

The most recent information on the relative prices of a wide range of treatment and disposal options for hazardous wastes is that prepared in 1980 by the Department of the Environment in their evidence to the House of Lords Select Committee. This information is shown in Table 7.

TABLE 7
1980 Prices for Hazardous Waste Disposal in the UK (1)

Treatment/disposal method	Waste type/ treatment method	Price range (£/tonne)
Landfill	Bulk solid wastes	2.5–10
	Bulk liquid wastes	5–30
	Drummed wastes	17.5–35 (3.5–7 per drum)
Physical treatment	Oil/water separation	10–15
	Sludge dewatering	5.5–12.5
	Solidification	12–100(2)
Chemical treatment	Neutralization of pickling acids	10–35
	Reduction of "typical" chromic acid waste	35
	Neutralization, precipitation and dewatering of a concentrated metal salt solution	50
	Oxidation of cyanide wastes	195–390
Incineration (land based) (3,4)	Bulk liquid wastes (high calorific value, low contamination) (4,5)	0–20
	Bulk aqueous waste (low contamination) (4,5)	50
	Bulk halogenated wastes	40–350
	Bulk or drummed solid wastes	45–900
Marine disposal	Disposal by discharge, from ships to coast wasters	2–5
	Marine incineration	65–95

(1) All these figures exclude the cost of transport which can be a large addition, ranging from £5–20 per tonne
(2) Chemical treatment of exceptional wastes can cost over £1000 per tonne
(3) Fluctuations in incineration costs are considerable and some energy rich waste even leaves the producer in credit
(4) Drummed liquid wastes are charged at higher rates, often at a premium per drum
(5) "Contamination" refers to halogen elements, sulphur, heavy metals, ash content

Since 1980, there have been considerable changes in the relative prices of different options, particularly those for landfill and incineration.

The Hazardous Wastes Inspectorate collected information on 1984 charges for landfill from a number of the larger waste disposal companies These are shown in Table 8, taken from the first HWI report. A feature of the table is the very significant discounts the contractors have had to offer major customers in order to preserve their market position.

The HWI comment on this table speaks for itself: "It is suggested that these prices are too low. The revenue earned at these prices will not be sufficient to meet the standards that must be achieved; allow proper restoration to be undertaken; provide a reasonable return of capital; nor secure the long-term future of companies involved in waste disposal."

While landfill prices appear to have declined significantly in real terms since 1980, the same is certainly not true of charges for incineration. Examples of 1986 prices include:

- PCB liquids: £1000 per tonne
- solids contaminated with PCBs: £1700 per tonne.

TABLE 8
Typical Landfill Disposal Charges: 1984

Waste type	List price	Discount price	Discount percentage (%)
Solid waste	2.50	1.50	40
Solid hazardous waste	5.00	2.80	44
Neutral sludges	14.00	7.00	50
Liquid hazardous waste	18.00	8.00	55

Crown Copyright. Reproduced with the permission of the Controller of HMSO.

ASSESSMENT

Since 1972, there have been significant steps forward in hazardous waste management in the UK. In particular the problems are now generally recognized and there has been substantial improvement in operational practices. In addition, a nucleus of an understanding of the problems now exists and there are now some very professional waste management companies.

Particular success stories underpinning this progress include general implementation of an adequate control system. Other highlights include the independent assessment of the situation by several influential bodies, including a Select Committee of the House of Lords and the Royal Commission on Environmental Pollution. Resulting from these assessments and other government activity have been the preparation of many technical memoranda and codes of practice for the management of particular types of hazardous waste.

We also can note that an active and continuing research program into the behaviour of hazardous waste, particularly in the landfill environment, has been in progress since 1972. Likewise, a major review of landfill practices has been published.

REMAINING PROBLEMS

The view of the Hazardous Wastes Inspectorate

Although much progress has been made, there is still considerable room for improvement. To quote a ministerial statement: "Higher standards of waste disposal can be achieved through better planning, greater professionalism, stricter policing and economic charging of waste producers."

The first two reports of the Hazardous Wastes Inspectorate (HWI) are certainly not complacent. Amongst their conclusions are the following:

- existing controls are very unevenly applied across the country;

- the standards of site licensing, inspection and enforcement vary widely between waste disposal authorities;

- this variation means that a few landfill sites in particular are able to accept waste that elsewhere is considered unsuitable for landfill, resulting in serious distortions in the market place and a downward pressure in prices;

- all of this is resulting in an evening of standards downwards; and

- the "cheapest tolerable means" is sometimes being adopted rather than the "best practicable means".

Most of these problems arise from two central principles of the UK system of control. These are devolution of control to local Waste Disposal Authorities, and the reliance on market forces.

It is apparent that some modifications to the way in which these principles are presently implemented is necessary in order to improve the operation of the system.

Devolution of control to local WDAs relies on uniformly high levels of enthusiasm and professionalism. Unfortunately, there are a large number of conflicting priorities within local authorities, and in many cases the waste disposal function is both understaffed and relatively low in the professional hierarchy.

The reliance on market forces for hazardous waste management depends on the existence of a consistent national strategy, so that waste is effectively directed to a facility representing the best practicable environment option (BPEO).

As a result of "unfair competition" from landfill, all of the existing merchant incineration and treatment facilities in the UK are on an economic knife edge. The regional facility serving Scotland and the North of England has already closed, and another such closure would mean that something in excess of 10,000 tonnes per year of the most intractable hazardous waste would have no BPEO for its treatment. In 1986, the HWI expressed doubts over the survival of some of the available capacity for chemical treatment "whilst landfill site licences permit concentrated and reactive residues to be deposited in bulk".

The heart of successful waste management is public confidence and acceptance. Landfill depends heavily on proper management and control, and the findings of the HWI that some operators are "downright neglectful" and "demonstrate either ignorance of or disregard for the technical guidance available", does not foster public confidence. Action to correct landfill site malpractices, and to raise the standard of the industry in accord with current best practice, is essential.

Public concern has recently been directed at the incineration of PCBs at two particular sites. This already has had repercussions elsewhere, an example being fierce public opposition to a proposal to store small quantities of PCB capacitors pending treatment. It is clear that any future proposals to build state-of-the-art hazardous waste treatment facilities will face strong local opposition.

The HWI is aiming to improve the quality and objectivity of its individual facility assessments through the use of risk assessment techniques. Environmental Resources Limited carried out a feasibility study in 1985, as a result of which a "hazard audit" system for risk assessment of co-disposal landfill sites is currently being developed.

Enforcement

One issue on which the early work of the Hazardous Wastes Inspectorate has succeeded in focusing attention is the importance of good enforcement, to ensure that proper standards are being maintained. The existence of a control system, however perfect it might appear on paper, is meaningless if it is not vigorously enforced.

To illustrate the level of commitment required, one of the former Metropolitan Counties in England employed 37 staff in their Pollution Control Division, with a total hazardous waste production of around 500,000 tonnes per year. About 24 of these staff were enforcement officers who worked a shift system, giving coverage from 06.30 to 22.00 on Mondays to Fridays and on normal hours at weekends. These officers visited all licensed storage, treatment and disposal sites, of which there were more than 200 in their area. Many of these were licensed only for construction waste, but it is not possible to ensure that hazardous waste is only going to facilities licensed to receive it unless one regularly inspects all sites where illegal disposal could be taking place.

The county had also instigated a running program to visit all waste producers, a massive task which was envisaged as operating on a 5-year cycle. The aim was to promote a two-way exchange of information, offering advice to producers and obtaining information on their waste arisings. The county strongly supported proposals for a register of waste producers, including a requirement to produce an annual return on waste arisings.

Two members of the pollution control staff were directly responsible for policing activities and co-ordinating prosecutions. This was in fact one of the few waste disposal authorities with a successful record of prosecuting offenders under the Control of Pollution Act.

Unfortunately, relatively few waste disposal authorities in the UK put this level of resources into enforcement activities. Successful prosecution of offenders currently depends upon actually witnessing an illegal deposition taking place, and this leads to a very cautious approach in many authorities regarding the bringing of prosecutions. Both the Joint Review Committee and the Royal Commission recently recommended that prosecutions should be permitted for breach of license conditions not related to the act of deposit itself and this is now being proposed.

Problem waste

Four particular types of waste are most often quoted in the UK as being problem waste. These are polychlorinated biphenlys (PCBs), asbestos, acid tars, and drummed waste.

Following the recent closure of the Scottish incinerator, there was a temporary shortage of incinerator capacity for PCBs, particularly capacitors and transformer carcasses. A combination of economic difficulties and public pressure could cause further closures.

There are problems both with the manner in which asbestos is landfilled, with the many hundreds of sites across the country in which asbestos is being placed, and with the very fact that asbestos is being landfilled at all. It could be argued that the best practicable environmental option for asbestos currently would be a few large regional depositories where operational standards could be properly supervised and where the sterilization of land would be concentrated in just a few controllable locations. In the longer-term, treatment of asbestos, for example by the Vitrifix process, could become practicable.

These intractable wastes, comprising a mixture of concentrated sulfuric acid with tarry organic compounds, arise from a number of chemical processes, including a re-refining of waste oil. Landfill as practised until recently did not constitute co-disposal, but rather entombment of highly reactive wastes in clay vaults. The new process mentioned previously together with the closure of one of the major waste producers, may have removed this problem.

The Hazardous Wastes Inspectorate would like to phase out the practice of drumming wastes prior to treatment or disposal. The LPRG report recommends that "as a general rule, direct landfilling of drummed waste is unacceptable". The problem with drums is that they are anonymous, and it is too easy to slip in a few drums that are not what they claim to be. Sampling and identification of all drums arriving at a landfill or treatment site is judged to be impracticable.

Wastes escaping control

It is obviously difficult to estimate the extent to which wastes are escaping control. However, the HWI have drawn attention to some problem areas.

Some "cowboy" operators do exist, particularly in major urban areas. Pollution control inspectors have been assaulted in the course of their duties. Particular attention is drawn to asbestos and the residues from small solvent recovery operations as wastes potentially evading controls.

Illegal "fly-tipping" of construction wastes is reaching epidemic proportions in some areas (due to local shortages of disposal sites and high transport costs). It is estimated that there are at least 150 incidents annually involving hazardous waste.

It is difficult to assess the extent of any problem associated with small generators of hazardous waste. In principle, there is no lower quantity limit associated with the consignment note system for special waste, so small generators should be treated in exactly the same manner as anyone else. Similarly, such waste should be directed to a site which is licensed to receive it.

Opinions vary as to how far this view corresponds with practice. It is undoubtedly possible to evade control within the law, for example by diluting small quantities of hazardous waste with other materials so as to come below the concentration limits under the special waste regulations, or by disposing of small quantities to sewer with similar dilution below discharge standards.

Future policies and expected developments

As outlined early in this report, various protracted reviews of hazardous waste management in the UK have recently been completed, and new regulations are either imminent or in the pipeline. Thus, in the short-term we can expect that:

- the Hazardous Wastes Inspectorate will continue to work towards an improvement in the standard of site licensing and a closing of the loopholes that variations in standards now produce;
- various loopholes in the scope of licensing provisions will be closed;
- a "duty of care" requirement will be placed on the waste producer to take all reasonable steps to ensure appropriate disposal of his waste;
- all waste handlers will need to register with the authorities;
- licensing regulations will be amended to allow the imposition of post-closure monitoring requirements at landfill sites; and
- charges will be introduced for site license holders, to fund the enforcement and control functions of Waste Disposal Authorities.

In the wake of the initial report of the Hazardous Waste Inspectorate, a number of more speculative predictions may also be made. Increased resources will be given to enforcement and site licenses gradually will be upgraded. Also, a co-ordinated national strategy for hazardous waste management may emerge. Finally, if a further merchant incinerator closes, then some involvement in the current free market system may be forced upon the government.

BIBLIOGRAPHY

The Control of Pollution (Special Waste) Regulations 1980. Statutory Instrument 1980 No. 1709, London: HMSO, 1980. Explanatory notes given in joint circular from the Department of the Environment (4/81) and the Welsh Office (8/81), London: HMSO, 1980.

The Control of Pollution (Collection and Disposal of Waste) Regulations 1987. Circulated for comment on January 16, 1987 by the Department of the Environment, available from Land Wastes Division at address in Ref. 6.

Department of the Environment, *Co-operative Programme of Research on the Behaviour of Hazardous Wastes in Landfill Sites.* HMSO, London, 1978.

Gray, D. A., Mather, J. D. and Harrison, I. B. *Quarterly Journal of Engineering Geology* 7, 1974, 191-96.

Hazardous Wastes Inspectorate, *Register of Facilities for the Disposal of Controlled Wastes in England and Wales, 1984*. Available from HWI, Department of the Environment, Romney House, 43 Marsham Street, London SW1P 3PY.

Hazardous Wastes Inspectorate, *First Report. Hazardous Waste Management — An Overview*. June 1985. Available from HWI at address in reference 6.

Hazardous Wastes Inspectorate, *Second Report. Hazardous Waste Management: Ramshackle and Antediluvian?* July 1986. Available as above.

House of Lords Select Committee on Science and Technology (1981), *Hazardous Waste Disposal*. Volume I — report. Volumes II and III — evidence, London: HMSO, 1981.

Roberts, D., "The Vitrifix Process". Proceedings of an International Conference, *Chemicals in the Environment*, Lisbon, July, 1-3, 1986. London: Selper Ltd.

Royal Commmission on Environmental Pollution; Eleventh Report, *Managing Waste: the Duty of Care*. London: HMSO, 1985.

Waldegrave, W. *Government Press for Higher Standards of Waste Disposal*. Department of the Environment press notice 142, March 27, 1985. Also in Environmental Data News Service (ENDS) Reports 123/April 1985, p. 9-11.

Waste Disposal Law Amendment, Consultation paper, September 15, 1986, available from Land Wastes Division at Department of the Environment, Romney House, 43 Marsham Street, London SW1P 3PY.

Wilson, D. C., The definition of special wastes. *Chemistry and Industry*, April 17, 1982, 253-259.

Wilson, D. C., Smith, E. T. and Pearce, K.W., Uncontrolled hazardous waste sites: a perspective of the problem in the UK. *Chemistry and Industry (London)*, January 3, 1981, 18-23

14

Hazardous Waste Management in THE UNITED STATES

WILLIAM S. FORESTER

ISW/ISWA Secretariat, American Public Works Association, Chicago, USA

OVERVIEW

More than 70% of the approximately 264 million tonnes of hazardous waste generated in the USA in 1981 was produced by the chemical and petroleum industries. For this reason, organic waste from these two industries are major problems. Of the remaining waste produced, 22% was generated by metal-related industries. This corrosive and metal-containing waste is likewise recognized as a major problem. Fortunately, these wastes are fully regulated under the nation's waste management law, the Resource Conservation and Recovery Act (RCRA) of 1976. With this law, and subsequent legislation, the legal framework is in place to bring hazardous waste under control.

There are, however, other problem waste that is either not controlled or only partially controlled. This includes waste oils, wastes from small hazardous waste generators, and waste from the mining industry. Amendments to RCRA dealing with the first two of these problem areas were enacted in November, 1984, and the regulatory programs to implement these amendments are being developed. Also, some waste still will not be fully regulated, or will escape regulation altogether. For example, small quantity generators, defined by the amendments as those generating between 100 kg and 1000 kg of hazardous wastes per month, will not be regulated as stringently as those producing above 1000 kg per month.

Generators producing less than 100 kg per month will not be regulated by Federal Law at all.

One disposal technology used extensively in the past that has produced major problems has been the practice of disposing of hazardous waste in landfills or other ground facilities such as pits, lagoons, and deep wells. This has resulted in land and ground water contamination in some areas, and in old or abandoned sites that threaten to pollute the environment and endanger human health. Also, additional control of underground storage tanks is planned. Fortunately, national laws have been enacted in the form of RCRA and the Comprehensive Environmental Response, Compensation, and Liability Act, "Superfund" legislation to help clean-up problem abandoned sites. In a separate area of concern, incineration of solid waste has resulted in the production of dioxin, which has closed or threatened to close several major resource recovery plants.

The US Environmental Protection Agency has compiled information on more than 25,000 sites where hazardous waste was managed prior to the establishment of regulations. Of these sites 888 have been designated as potentially in need of remedial actions and have been placed on a Natural Priority List. This "national" list was compiled by EPA with the support and recommendations of the 50 states. These sites are eligible for clean-up utilizing a 9 billion dollar trust fund under Superfund legislation amended in 1986. Since the start of this program detailed investigation and planning for remedial work has begun on 473 material priority sites, remedial action designs have been completed at 110 sites and remedial actions have been carried out at 143 sites (as of October 1986)

Standard practices for detoxifying chemicals, neutralizing strong acids and bases, and treating corrosives, all have long been available for use in the USA. Refinements of these practices continue. Also, several new high temperature incinerators have been built, including three large ones for PCBs and other hazardous organics. The US EPA has developed and put into use a mobile high-temperature incinerator for on-site clean-ups. The legislation in the USA is adequate, with the major RCRA amendments of 1984, to support an effective regulatory program. Technical standards have been developed for treatment facilities, new landfills, and a national manifest system.

Priorities for national action center around those provisions of the 1984 amendments to RCRA and the 1986 amendments to superfund. These RCRA amendments will have a profound effect on the management of hazardous wastes in the USA, in that new regulatory programs must be developed to achieve the following four waste management goals:

- The prohibition of disposal of hazardous waste on the land.

- The development of more stringent standards for landfills and waste lagoons.

- The regulation of small quantity generators, or those generators producing between 100 and 1000 kg of hazardous waste per month.
- The control of underground tanks that store petroleum products and hazardous substances.

The last provision entails a major expansion of EPA's regulatory responsibility. For the first time EPA will be regulating hazardous materials as well as hazardous waste.

An additional priority area concerns Superfund. A new appropriation of funds, of the magnitude of 9 billion dollars over a 5-year period is expected to greatly accelerate the clean-up of old problem sites.

The 1986 amendments to Superfund also include:

- Increased emphasis on the use of alternative treatment and recycling technologies. Technologies that reduce volume, toxicity and mobility of hazardous waste are to be preferred.
- New focus on cleaning up leaks from underground storage tanks containing petroleum and hazardous substances. Up to 500 million dollars of the trust fund can be used for this purpose.
- A special set of provisions dealing with prevention and mitigation of the accidental release of hazardous substances (such as occurred in the tragedy in Bhopal, India). The amendements require emergency planning, notification, reporting and an emissions inventory.

NATIONAL CONTROL SYSTEMS

Summary of legislation

Federal law in the USA designates the US Environmental Protection Agency as the authority responsible for developing and enforcing programs to ensure the safe management of hazardous waste. This responsibility is mandated basically in the form of two far-reaching laws: The Resource Conservation and Recovery Act of 1976 (RCRA); and the Comprehensive Environmental Response, Compensation, and Liability Act of 1980 (CERCLA or Superfund).

Under RCRA, the EPA is charged with establishing controls to prevent present and future threats to human health and to the environment. Under CERCLA, EPA has the authority to clean-up old or abandoned hazardous waste problem sites. CERCLA provides the funds to clean-up these sites, if direct generator responsibility cannot be readily established. Where waste ownership can be established, fines or clean-up costs can be directly assessed to those responsible.

In November, 1984, the Resource Conservation and Recovery Act was amended, in the form of the Hazardous and Solid Waste Amendments, to place additional restrictions on the land treatment of hazardous waste, to require the regulation of small quantity generators, and to place restrictions on underground storage tanks. In 1986 the Superfund legislation was also amended.

History of hazardous waste management

In relatively small quantities, hazardous waste as a by-product of human industrial and commercial activities has existed in the USA since the country's beginning in 1776. It has been since World War II and subsequent post-war industrial expansion, however, that the hazardous waste problem has surfaced as the pervasive one we now face. The early reaction to getting rid of hazardous waste was to dump the waste in the national waterways.

The beginning of the end of this practice came in Pennsylvania in the 1940s when that state began enforcing provisions of its Clean Streams Act. Many industrial plants then began the practice of storing waste in earthen lagoons on plant property. Deep well injection also came into use, as did the common practice of landfilling hazardous waste. Publicity from such problem sites as Love Canal in New York, the "Valley of the Drums" in Kentucky, major spills in several areas, and major problem fires in New Jersey and other states, forced public attention on the problems of indiscriminate disposal.

Beginning with RCRA in 1976, strong national controls were put in place to prevent indiscriminate disposal harmful to the environment and human health. CERCLA was enacted in 1980 to deal with the consequences of past unsound disposal practices. Major amendments to RCRA and CERCLA were enacted in late 1984 and 1986 respectively to deal even more comprehensively with hazardous waste problems.

Licenses and manifest systems

A program of permits for treatment, storage, and disposal facilities was established with the passage of RCRA. The permit must contain specific requirements for the operation of the facility. To avoid the shutdown of a facility until a permit can be issued, RCRA allows a facility that was in existence as of November 19, 1980, to operate under "interim status", if the facility meets certain requirements including the submission of an application for a full operating permit.

To support this permitting system, EPA has published regulations governing the location, design, and construction of facilities. The regulations require disposal facilities to conduct groundwater monitoring to detect leakage of contaminants. They also require new facilities to meet design standards and location requirements.

While transporters are not licensed under the national regulatory system, RCRA does establish a manifest system to tract hazardous waste from generation to ultimate disposal. The law requires that transported waste be accompanied by a manifest that specifies the origin, type, and quantity of waste, as well as the facility designated to receive the waste. The receiving facility must return a copy of the manifest to the generator. The manifest system is intended to ensure that hazardous waste is disposed of only at approved treatment, storage, or disposal facilities. EPA published in March, 1984, the actual manifest form that must be used for all shipments. Thus, the system is basically uniform throughout the 50 states.

Roles of national, regional and local governments

State governments with staffs and technical capabilities sufficient to enforce the mandates of the national law may, under RCRA, assume operation of the state's hazardous waste regulatory program. When a state is authorized to operate the program, the federal government provides guidance and oversight, and has back-up enforcement responsibility. States may be authorized for all or certain categories of a hazardous waste program. EPA may authorize a state program only if it is "equivalent to" the federal program, is "consistent with" the federal program, and provides adequate enforcement. EPA also provides financial support to state hazardous waste programs through matching grants.

Regional authorities and districts are set up in areas throughout the USA to deal with environmental management problems peculiar to individual river or major lake basins, or other ecological systems. These authorities can have impact upon governmental and industrial activities through the publication of policies, reports, and studies, but they tend not to have regulatory authority. Local governments may have regulatory authority where local laws and ordinances have been enacted granting them this power. In many instances local governments have enacted regulations more stringent or more comprehensive than the national regulations. Many local governments have extensive capability in the handling of spills and other hazardous waste emergencies.

Subsidies

Subsidies for assistance with hazardous waste regulation exist in the form of matching federal grants directly to state agencies charged with hazardous waste regulatory responsibility. These grants are authorized under Subtitle C of RCRA. They are managed through the US EPA's 10 regional offices. In fiscal year 1985, $56.7 million was awarded to the 50 states for this purpose. The individual grants varied from $4,483,000 awarded to Texas, to $85,500 awarded to Hawaii. An estimated $65 million was awarded in fiscal year 1986. This increase in funding over 1985 included monies to assist with management of the new underground storage tank regulatory program.

Transport systems

The hazardous waste transport system in the USA is basically a private sector system, with the government acting as the regulator through the national manifest system. Of the approximately 264 million metric tonnes of hazardous waste produced in the USA in 1981, 96% was managed on-site. Thus, only 4% was shipped to disposal sites, mostly by truck, and to a lesser extent by rail. The transporter is not licensed, but must be able to demonstrate capability to the generator, and must be knowledgeable enough about the trade to participate in the manifest system. Also, US Department of Transportation regulations concerning placards and labeling and packaging must be followed, as must local transport restrictions.

Responsibilities

Generators of hazardous waste are responsible for evaluating their waste to determine if it is regulated under RCRA. If their waste is transported, generators must prepare the manifest and must package and label the wastes according to EPA and US Department of Transportation regulations. They also must maintain certain records, and send periodic reports to EPA. Under the manifest system, the generator is primarily responsible for determining how its waste is to be disposed of, and for ensuring that its waste is received by the intended treatment, storage, or disposal facility.

A transporter may only accept a waste that is accompanied by a manifest, and must transport the waste only to the facility identified on the manifest. If a waste is spilled during transport, the transporter must take immediate action to protect human health and the environment.

Treatment, storage, and disposal facilities may operate only under interim status, if final approval of their permit applications has not been granted. This includes most facilities. Under interim status, facilities must meet certain requirements including the preparation of a waste analysis plan, personnel training, utilization of the manifest system, ground-water monitoring, closure and post-closure plans, and financial responsibility. Final permits include more extensive ground-water monitoring requirements, and other design, operational, and reporting requirements.

Programs for dealing with old, closed or abandoned sites

While RCRA deals primarily with the regulation of treatment, storage, and disposal facilities, the Comprehensive Environmental Response, Compensation, and Liability Act (CERCLA, or Superfund), as stated previously, gives the US Government authority to clean-up old, closed, or abandoned sites. Specifically, the federal government is authorized to respond directly to releases, or threatened releases, of hazardous substances that may endanger public health or welfare.

Clean-ups are paid for from a 9 billion dollar national fund. The source of this fund is as follows: 2.75 billion dollars from a petroleum tax, 1.4 billion dollars from a tax on chemical feedstocks, 2.5 billion dollars from a broad-based tax on corporations, 1.25 billion dollars to be appropriated from general revenues, 300 million dollars interest, 300 million dollars to be recovered from private parties and 500 million dollars from a tax on fuel stored in underground tanks. The fund is reimbursable in that the government generally can take legal action to recover its clean-up costs from those subsequently identified as responsible for the release. Anyone so identified as liable is subject to punitive fines equal to three times the government's clean-up costs.

A "National Contingency Plan" spells out the guidelines and procedures the federal government will follow in implementing the Superfund law. While EPA's enforcement office seeks to identify those responsible and require them to pay for clean-up costs, EPA can take direct action in the following three ways when such identifications can not be made:

- Immediate Removals. Immediate removals may be undertaken when a prompt response is needed to prevent harm to public health, welfare, or the environment. Immediate removals may be necessary to avert fires or explosions, to prevent exposure to acutely toxic substances, or to protect a drinking water source. Actions that the EPA may take are the installation of security fencing, the construction of physical barriers to control discharges, or removal of hazardous substances from the site. There are limits to the use of this type of removal response. Ordinarily responses must be made with 6 months of identifying the problem and must not cost more than 1 million dollars.

- Planned Removals. Planned removals may be undertaken when an expedited, but not necessarily immediate, response is required. Such actions are intended to minimize harm that would occur from exposure if action were delayed. These removals are subject to the same time and cost limits as immediate removals.

- Remedial Actions. Remedial actions are longer term, usually more expensive, and aimed at permanent remedies. For this reason they may be taken only at sites listed as national priorities. Specific on-site actions may include removal of drums containing waste, caping the installation with clay or other material, construction of dikes or ditches to control surface or drain water, liners, and grouting to control ground water. Off-site, provisions may be made for alternate water supplies if existing supplies have been contaminated, and for the relocation of residents if endangerment warrants.

The primary responsibility for managing the Superfund program was assigned by Executive Order to EPA. Related responsibilities, however, have been assigned to other federal and state agencies. The US Coast Guard has primary responsibility for spills that occur in coastal areas. Under agreement with the federal government, state governments may assume responsibility for planning and managing response programs. Usually, private contractors perform the actual work at a site, under federal or state government supervision.

Superfund monies may be spent only after careful evaluations have been made as to responsibility and the funding available from those identified as responsible. The following limitations exist:

- A response financed by Superfund may not be taken if EPA determines that the owner, operator, or other responsible party, is undertaking an appropriate clean-up.

- Immediate removals may be taken only to bring a release of hazardous substances under control. This category of response was not devised primarily to eliminate long-term problems.

- The state in which the problem site occurs must agree to pay a share of the project's cost before a remedial action or planned removal can be taken. The state also must agree to maintain the site after response work is completed, and must provide for off-site disposal, if necessary.

- Superfund monies may not be used in specific situations where the potentially harmful substances are covered by other laws. An example of this exclusion is nuclear material from a nuclear incident.

Guidelines for determining the extent of clean-up appropriate, and the most cost-effective expenditure of funds, as stated above, are found in the National Contingency Plan. A major requirement is that the costly step of excavating hazardous wastes and transporting them off-site for disposal be avoided whenever possible. Also, benefits of cleaning up one site must be weighed against benefits of cleaning up other sites elsewhere in the nation. Hard decisions must be made, and funds shifted to where they can be most effective.

DEFINITIONS OF HAZARDOUS WASTE

In the USA, the purpose of defining hazardous waste is to identify as much waste as possible that is harmful to the environment and human health, but to accomplish this under a system that is simple to understand and primarily self-implementing. The procedures for defining hazardous waste are set forth under RCRA. Specifically, it defines a hazardous waste as a "solid waste" that may pose a hazard when "improperly treated, stored, transported, or disposed of, or otherwise mismanaged". RCRA further defines a solid waste to include liquid, semi-solid, or gaseous materials. A waste is determined to be hazardous if it is specifically listed by EPA regulation, or if it meets one or more of four basic criteria: ignitability, corrosivity, reactivity, or toxicity.

The implications of the hazardous waste definitions for the National Control System are far-reaching. The estimated 264 million tonnes under regulation involves more than 14,000 generators. Wastes are regulated from industrial, commercial, mining, agricultural, and community sources. Several hundred specific hazardous wastes are listed in regulations promulgated by EPA. These include waste solvents, waste from inorganic or organic chemical production, and discarded chemicals and products.

If a waste is defined as hazardous, the waste is controlled under RCRA's "cradle-to-grave" management system. Some wastes exhibiting a characteristic, however, are specifically excluded from control under RCRA. These include hazardous waste treated exclusively in waste water treatment tanks regulated under the national water quality law's National Pollutant Discharge Elimination System (NPDES) permit program, waste generated in conjunction with ore and minerals extraction and benefication, waste legitimately disposed of through sewers to publicly owned waste water treatment works, household hazardous waste, and small quantity generators of hazardous waste generating less than 100 kg per month.

Also, there is considerable range in the preciseness of the definitions of the four characteristics, thus affecting what the definition covers. The toxicity characteristic, employing a well-defined standardized test, is the most precise. At the other end of the scale is the reactivity characteristic, which is included in the form of a narrative definition. Current effort at the US EPA seeks to produce a more precise definition for the characteristics of ignitability, corrosivity, and reactivity. Additionally, the toxicity characteristic will be broadened to include more than the current six organic compounds. When this work is completed, many more wastes will be regulated under the RCRA system.

Under the RCRA system, if a waste is not defined as hazardous, it is not regulated as such under Federal Law (RCRA). The generator however does have to comply with other federal or state law. The NPDES permitting program, mandated under the Water Pollution Control Act, is an example of this. Municipal solid waste is not regulated as hazardous waste. There are a set of Federal criteria covering disposal of municipal solid waste, but they are enforced primarily by state or local agencies.

SOURCES AND QUANTITIES OF HAZARDOUS WASTE

By far the best and most comprehensive information on hazardous waste is compiled on an on-going basis by the US EPA. In a national survey in 1981, EPA estimated that 14,100 generators produced approximately 264 million tonnes of regulated hazardous waste. Most of those generators, 84%, shipped all or part of their waste to another site for treatment, storage or disposal. Only 16% of the generators managed their waste at the site of generation. In terms of quantity, however, it was found that 96% of the waste was managed on-site, while only 4% was shipped off-site. Only 9% of the hazardous waste generated in 1981 was recycled at the site of generation. Approximately 4800 facilities treated, stored, or disposed of hazardous waste in 1981.

Table 1 presents the quantities of hazardous waste generated in 1981 by industry type. Manufacturing was responsible for producing more than 90% of the total quantity of hazardous waste generated, or 245 million tonnes. The chemical and petroleum industries alone produced more than 70% of the total generated, or 187 million tonnes. The metal-related industries generated 22% of the total, 58 million tonnes. All remaining manufacturing and non-manufacturing industries, and industries not specified by kind, accounted for about 7%, or 19 million tonnes of hazardous waste generated in the USA.

TABLE 1
Quantities of Hazardous Waste Generated in the USA in 1981 by Industry Type

Industry type	Quantity (million tonnes)
Chemical and petroleum	187
Metal-related industries	58
Other industries	19
Total	264

Table 2 presents by waste group type the quantities of hazardous waste handled by management facilities in 1981. Caution should be exercised in using this information since a high degree of uncertainty is associated with the quantities shown. This is due to sampling and non-sampling errors and to other problems associated with gathering the information. In Table 2, the sum of the amounts handled by the different waste groups exceeds the 264 million tonnes generated/managed in the USA. This difference is due to the various interpretations of "handled" by survey respondents, and to the fact that some respondents reported hazardous waste quantities under multiple hazardous waste code groups. As a result, some quantities were added to more than one of the categories presented in the table.

TABLE 2
Quantity of Hazardous Waste Handled by Management Facilities in the USA in 1981 by Type of Waste Group

Type of waste group	Quantity (million tonnes)
Spent halogenated and non-halogenated solvents	11.9
Electroplating and coating waste water, treatment sludges and cyanide-bearing bath solutions and sludges	9.7
Listed industry wastes from specific sources	48.4
Off-specification or discarded commercial chemical products and manufacturing intermediates	10.8
Acutely hazardous waste	0.7
Ignitable waste	5.2
Corrosive waste	122.7
Reactive waste	11.9
EP toxic waste	41.3
Unspecified (including state-regulated and self-defined hazardous waste)	43.5
Total	306.1

COLLECTION AND TRANSPORT SYSTEMS

Controls for the collection and transport of hazardous waste come from two sources in the USA. These are in the form of manifest requirements set by the US EPA, and in the form of rules for placarding, labeling, and containerizing transported hazardous wastes set by the US Department of Transportation (DOT).

The manifest system for tracking hazardous wastes from generation to ultimate disposal is an essential part of the hazardous waste management system in the USA. RCRA places responsibility on the generator to prepare a manifest, properly package and label the waste, and ensure that the waste is received by the facility to which it is sent. The transporter cannot accept waste if it is not accompanied by a proper manifest.

The manifest form must accompany all waste during shipments. The receiving treatment, storage, or disposal facility must return a copy of the manifest to the generator. If a generator does not receive the return copy within 35 days, the generator must contact the receiving facility. The generator must notify EPA if verification that the waste was received at the facility is not received within 45 days.

Hazardous waste transporters are not licensed by EPA. However, they must receive an EPA Identification Number. This number is placed on the manifest. If a spill occurs during delivery, the transporter must take immediate action to protect human health and the environment. Generators and transporters must keep copies of the manifest for 3 years. To avoid duplicate requirements, RCRA regulations have incorporated US DOT regulations governing transport of hazardous materials. Initial regulations allowed states to develop their own manifest forms. Because of confusion, however, EPA and DOT now require the use of a Uniform Hazardous Waste Manifest form.

Transporters of hazardous waste, as stated above, are not licensed in the USA. For control purposes, however, transporters must have an EPA identification number, and must participate in the national uniform manifest system. Most transporters of hazardous wastes in the USA are private firms.

Waste Oil Collection Systems

The US EPA has proposed administrative requirements that will create a system for tracking used oil from the initial marketer through the distributor to the disposer (burner). This will, in effect, bring the collection, transport, treatment, and disposal of used oil under one comprehensive national control system.

Used oil, while not regulated as a hazardous waste *per se*, may be so regulated if it contains substances regulated under RCRA as hazardous waste or if it meets the hazardous properties criteria defined in the statute. In these cases the collected material is subjected to regular manifest requirements.

Currently, about 4.5 billion liters (1.2 billion gallons) of waste oil are generated in the USA each year by several hundred thousand waste oil generators. In most cases, a collector regularly visits each generator and pumps accumulated waste oil into a tank truck. Some collectors segregate crank case oils from industrial oils because of their different compositions.

Some states have implemented programs for monitoring waste oil transactions, but most of these programs likewise are still in the early stages of development. As a result, much of the collection, reprocessing, and reuse of waste oils in the USA is not documented. We do know that the ultimate disposition of the collected oil in most geographic regions is dictated by market conditions. For example, a collector who oils roads in the summer may also reprocess or blend used oils into boiler fuels during the winter months when the road oiling market is slow and heating needs are great.

Household hazardous waste collection systems

Historically, household hazardous waste has not been regulated nationally in the USA. There are numerous programs at the local level, however, for the special collection and disposal of these wastes. Likewise, many communities publish guidelines for households to follow in the special management of waste hazardous products. Successful programs have been developed for Dallas, Texas, San Diego and Los Angeles, California, and many other communities. The Golden Empire Health Planning Center, Sacramento, California, has published a handbook on the subject for local communities, "Household Hazardous Waste Solving the Disposal Dilemma". The US EPA is developing a program to work directly with local governments and groups in developing control programs.

STORAGE, TREATMENT, AND DISPOSAL SYSTEMS

Storage systems

The three methods used most frequently in the USA for hazardous waste storage are in containers, in tanks, and in surface impoundments. Containers are defined by RCRA regulations as, "any portable device in which material is stored, transported, treated, disposed of, or otherwise handled". Fifty-five gallon metal drums are the most commonly used hazardous waste containers in the USA. This is followed in order of usage by 25 gallon plastic drums, tank cars, tank trucks, and, for some acutely hazardous wastes, glass vessels.

According to the 1981 national survey, approximately 85% of the 4300 identified hazardous waste storage facilities stored waste in drum and vessel containers. Tanks were the second most commonly used hazardous waste storage container, used by 33.2% of the 4300 hazardous waste storage facilities during 1981. About one-eighth of the 4300 storage facilities, or 550 facilities, stored hazardous waste in surface impoundments during 1981.

Although surface impoundments are only the third most frequently used storage method in the USA because of their large capacities they account for a large proportion of the hazardous wastes stored. Surface impoundments in 1981 were used to store an estimated 52 million tonnes of hazardous waste, or about 39% of the total quantity estimated to be stored. Tanks were used to store the second largest quantity of hazardous waste, about 19 million tonnes. Storage containers were used to store about 600,000 tonnes of hazardous waste. The majority of the remainder of the 135 million tonnes was not allocated by the survey to a particular storage method.

Incineration systems

Large amounts of hazardous waste are incinerated in the USA in facilities designed for that purpose. Incinerators are defined by RCRA regulations as, "an enclosed device using controlled flame combustion, the primary purpose of which is to thermally break down hazardous wastes. Examples of incinerators are rotary kiln, fluidized bed, and liquid injection incinerators." All of these design types are currently in use in the USA.

According to the most recent figures from EPA, there are approximately 230 hazardous waste incinerators in service. Most of these are liquid injection units but approximately 40 are rotary kilns. The exact number of cement kilns and boilers that have been used to burn hazardous waste is not known, although a few cement kilns are currently accepting almost all kinds of hazardous waste. A number of fluidized-bed incinerators are in service, and limited burns of an at-sea liquid injection incinerator system have been conducted. The EPA recently proposed regulations defining performance characteristics, operating practices, and permit requirements for ocean incineration.

The Hazardous and Solid Waste Amendments of 1984 established more stringent requirements for the burning of hazardous waste, and EPA has promulgated regulations that distinguish between incineration and the burning of wastes to recovery energy. EPA estimated that approximately 1.7 million metric tonnes of hazardous waste were treated in incinerators in 1981.

Landfills and disposal of waste on the land

In the 1981 EPA study, it was found that 200 landfills nationwide disposed of approximately 3.0 million tonnes of hazardous waste. Regulatory requirements mandate stringent controls for these facilities. First, all landfills, as well as all other land disposal facilities, are required to have permits that define specific design and operating requirements for the facility. For landfills and all other facilities that dispose of hazardous waste on the land, the permits specify requirements for liner design, liner thickness and composition, leachate collection and treatment systems, and caps to prevent infiltration of precipitation.

Groundwater monitoring is required for at least 30 years after a facility is closed. If groundwater is contaminated by the facility, the permit specifies the corrective actions necessary to contain, remove, and treat the groundwater.

The Hazardous and Solid Waste Amendments of 1984 recognize that all land disposal methods could potentially release hazardous materials into the environment. Thus, the amendments include a set of provisions that will severely restrict the disposal of waste on the land in the future. They require the EPA to evaluate each hazardous waste identified in its regulations and decide whether those wastes can be safely disposed of on the land.

The act states that "the land disposal of hazardous waste is prohibited unless EPA determines that such a prohibition is not required in order to protect human health and the environment for as long as the waste remains hazardous". The act sets forth a very aggressive schedule for EPA to review all hazardous waste for this purpose. If EPA fails to make a decision within the allotted time, the prohibition on disposal is put in place automatically by the new law. This clearly demonstrates the strength of the inclination against land disposal contained in the act. Unless EPA positively concludes that the land disposal of a hazardous waste can be accomplished safely, that disposal is prohibited by specific legislative "hammer" or mandate.

In order to carry out the prohibition of the land disposal of hazardous waste as specified in the amendments, it may be necessary in some instances to allow time for alternate treatment and destruction capabilities to become available. For this reason, the act allows strict prohibition against land disposal to be delayed for the time necessary to construct alternative systems to treat or destroy the waste. Taken together, these new more stringent standards for land disposal will drastically change the way hazardous waste are disposed of in the USA. They will increase the costs of land disposal and create further impetus for the reduction of waste generated, recycled, and treated in alternative ways.

This prohibition against landfilling and other forms of land disposal does not, however, apply to all types of hazardous waste. Thus, some low-level hazardous waste will continue to be disposed of in landfills, and treated, stored, or disposed of in surface impoundments (lagoons). This waste will include residues from treatment systems, incinerator ash, neutralized waste, solidified waste, and low mobility waste with low levels of toxic constituents. Facilities managing this waste will have to comply with very stringent design and operating standards.

Underground disposal in mines or caves has only been used to a very minor extent in the past, and with the new amendments limiting land disposal of hazardous waste it most likely will be used even less in the future. In large part, this disposal method is limited to certain geographical areas, where geological conditions such as groundwater seepage and subsequent contamination would not occur. The 1981 report showed mine and cave disposal to be an almost insignificant amount. This procedure would be viewed more as a temporary storage procedure in the USA than as an acceptable disposal practice.

The 1981 EPA report showed that 90 injection wells nationally disposed of approximately 32 million tonnes of hazardous wastes. This disposal option will be restricted when the 1984 RCRA amendments come fully into effect.

The 1981 EPA report listed 116 surface impoundments, which were used to dispose of 19 million tonnes of hazardous waste. The RCRA amendments essentially will place the same regulatory limits upon this type of land disposal as upon landfills.

Co-disposal is essentially not used as a disposal method in the USA. Where mixing of waste does occur, the resulting waste is almost always regarded as hazardous.

The 1981 EPA report listed 70 land treatment units in the USA. These units accounted, however, for only 400,000 tonnes of hazardous waste, a fraction of a percent of the total disposed. Again, the RCRA amendments will place this disposal option under strict regulatory control.

Hazardous waste is not allowed to be disposed of in the ocean coastal areas surrounding the USA. However, waste waters and other non-hazardous waste media such as sewage sludge and dredge material are disposed of in ocean areas. The areas involved most heavily are the major urban/industrial areas: Boston Harbor, the New York Bight, the Biscayne Aquifer in Florida, the New Orleans Delta and the East Texas Coast, the San Diego/Los Angeles area, San Francisco Bay, and Puget Sound. Data in all of these areas indicate that toxic metallic and organic substances have contaminated the waters, the sediments, and aquatic wildlife.

A case in point is the New York Bight area of the Coast of New York City and industrial northern New Jersey. In 1983, 8.3 million wet tonnes of sewage sludge were dumped from barges at a site 12 miles from the coast, 4.1 million cubic yards of sand and silt dredged from New York Harbor were dumped at a site even closer to the coast, and 38,000 tonnes of acid waste were dumped at a third site southeast of the sewage disposal site. Tests consistently show that sewage sludge from USA municipalities contains viruses, bacteria, trace metals, organic substances, organic chemicals, metals, and oils. The 12 mile site has been a significant source of contamination to the coastal waters of New York and New Jersey, according to both federal and state studies. The dredged material, while at times heavily contaminated, tends to sink quickly to the bottom and remain at the disposal site. The acid that is disposed of in the bight area is thought to be neutralized by the salt water.

The US EPA recently has acted under provisions of the Water Pollution Control Act as well as RCRA to place controls on dumping in the New York Bight area. In May, 1984, EPA announced plans for an interim decision that would move the sludge disposal site to a location 106 miles southeast of New York Harbor. What disposal is taking place currently at the 12 mile site is done on the basis of court orders favoring petition from New York and New Jersey to allow them to continue to use the site. Dredged materials and acids will continue for the time being to be dumped at their sites in the Bight area. The EPA, however, has placed volumetric limits on dredged material dumping, and will issue new permits for acid disposal for a site in the 106 mile area, phasing out the site nearer the coast. The EPA will eventually phase out all dumping in the New York Bight.

Energy recovery and recycling

According to the 1981 national survey, approximately 4% of the RCRA regulated hazardous wastes, amounting to approximately 2.8 billion gallons, was recycled. Of this amount, more than 80% of the waste was recycled on-site while the remainder was recycled off-site. Generally speaking, recycled waste is regulated under RCRA as hazardous waste if recycled in a manner that constitutes disposal. Examples are:

- wastes applied to or placed on the land in a manner that constitutes disposal,
- waste burned to recover energy or used to produce a fuel, or
- waste accumulated for later resell or reuse, unless 75% of the accumulated waste is recycled during the following 1-year period.

Normally, use and reuse is not considered waste management, and this waste would not be regulated.

The recycling of waste by burning is a common practice in the USA, and one that is coming under increasing scrutiny and regulation. Currently, about 230 million gallons per year of hazardous waste are burned for energy recovery in approximately 1450 boilers and industrial furnaces. These wastes are primarily spent solvents. Another practice that will soon be regulated is the burning of recycled waste oil in commercial and large apartment boilers. The EPA will soon establish maximum concentration limits for certain inorganic and organic constituents for waste oil recycled into new products.

Description of programs to reduce waste generation

Hazardous waste reduction is usually defined as source reduction, which includes process modification and substitution of less hazardous materials for ones currently in use. In addition, waste reduction can encompass on-site recycling and treatment.

A Treatment, Recycling, and Reduction Program has been created within the US EPA to study various treatment technologies. The program is specifically directed toward gathering and disseminating technical information on currently available or emerging treatment and recovery technology and on existing treatment capacity. A regulatory framework will also be established to set treatment standards and compliance dates, if needed, to control certain chemicals.

The major treatment technologies that are currently being examined with regard to their applicability to specific waste streams include:

- Biological treatment processes.
- Dechlorination to detoxify persistent chlorinated substances.

- Carbon absorption, a technique that has proven particularly useful for removing organic compounds from waste water.
- High temperature incineration.
- Neutralization of wastes.
- Chemical oxidation of primarily organic materials such as cyanides, phenols, and organic sulfur compounds.
- Precipitation.
- Solidification and stabilization.

Initial activities are focused on technologies available to treat the specific waste most likely to be the first banned from land disposal. Among the present candidates are wastes containing solvents, dioxins, metals, halogenated organic compounds, corrosives, cyanides, and other reactives. Reports on these wastes and on technologies available to treat them are currently being completed. Additional studies addressing the costs, effectiveness, development time, and the current capacity of specific treatment technologies, is set to begin soon.

In addition to technical and regulatory efforts, a three-way information program will be undertaken with industry, hazardous waste regulating agencies, and the public, to exchange information about EPA's alternative technology program and the results of technical studies. Field measurements and demonstration projects will also start this year to evaluate the applicability of certain technologies to particular wastes. These studies are needed to fill data gaps in the existing technical literature and will be conducted primarily by EPA's Office of Research and Development.

Existing research suggests that already known alternative technologies can be used or are already being used to treat and recover various kinds of hazardous waste. Many solvents, for example, are already being recycled or incinerated. A major problem with most alternatives is that they currently cost more than land disposal.

Operation and ownership of facilities

The large majority of the hazardous waste disposal facilities in the USA are privately owned. Most treatment facilities are co-located on-site with the processes that generate the wastes, and are owned by the firms producing the wastes. Of the approximately 4800 facilities that treated, stored, or disposed of hazardous waste in 1981, EPA estimated that there were about 326 "commercial" waste disposal facilities, a commercial facility being defined as a facility that receives more than half of its hazardous wastes from other firms. These commercial facilities managed approximately 4.8 million tonnes of hazardous waste in 1981, or 2% of the total.

Imports and exports

Current EPA regulations allow the import and export of most types of hazardous waste, although very little quantitative data exists concerning the exact amount of waste imported or exported by the USA. The information that does exist indicates that the USA imports very little hazardous waste from other countries.

The USA does export small amounts (compared to the total generated) of hazardous waste to neighboring countries (Canada and Mexico), with small amounts also going to the UK and the Federal Republic of Germany (FRG). The import and export of any waste containing PCBs at concentrations of 50 ppm or greater is forbidden under the US Toxic Substance Control Act.

ASSESSMENT

Hazardous waste enforcement in the USA, because of new and amended legislation, is in a state of flux. Enforcement has been successful in areas where the original 1976 law was specific and where regulations have been developed, but enforcement has not been comprehensive due to the original law not covering all aspects of hazardous waste management. With the 1984 Amendments to RCRA, however, sufficient laws currently are in place to provide the framework for an effective national enforcement program.

The control system implementation parallels the implementation of the enforcement program. The need for new control systems has been identified. They are currently under development. The general thrust of this development is that more waste will be classified as hazardous under RCRA. Likewise, the disposal of certain waste on the land will be prohibited, and standards for land disposal facilities will become more stringent. CERCLA has been expanded to provide additional funding to clean-up problem sites.

The three major sources of unregulated hazardous waste at present are: generators producing less than 100 kg per month, household hazardous waste disposed of in municipal waste streams, and those wastes that may be hazardous but have not yet been identified as such and listed in EPA regulations.

The EPA is has begun programs to regulate small quantity generators, identified as those who produce between 100 and 1000 kg per month. This includes such generators as automotive service stations, dry cleaning firms, and other small producers. Underground tanks containing hazardous substances and petroleum products also will come under control in the near future. In other programs, regulations will be developed to control the use of hazardous wastes as fuels, and to control other waste recycling activities.

Indications are that the current control system is working. The result of the strong requirement placed upon the disposal of hazardous waste has been to eliminate the smaller disposer less capable of meeting federal guarantees for human and environmental protection. Hazardous waste handlers were required to certify to the EPA by November 8, 1985, that they could meet federal requirements to monitor ground water for contaminants, and to obtain insurance to cover the cost of cleaning up any toxic leak that might occur at their sites. Only 500 of the nation's 1600 hazardous waste treatment, storage, and disposal facilities were able to make these certifications. The result is that the larger commercial disposal facilities will stay in business, with the smaller disposer shutting down. The expectation is that those firms that do remain in operation will be better able to meet the mandates of a strong national control system than was the complex of firms that existed before controls were put in place.

Expanded action is under way in the Superfund area, following the 1986 Congressional reauthorization of CERCLA. Related to this program, the EPA will complete its inventory of sites where hazardous substances and waste have been discharged. This inventory includes approximately 25,000 sites. The National Priority List will be increased to include between 1500 and 2000 sites. Enforcement actions under both statutes are expected to increase in order to bring regulated industries into compliance with the RCRA standards, and to force responsible parties to clean-up hazardous discharges under CERCLA.

Things being done well include: the development of hazardous waste control legislation, which is adequate to support effective regulatory program; the appropriation of public funds under the Superfund program to support the start of clean-up of old abandoned sites; the development of a national inventory to identify problem sites for Superfund clean-up; the development of technical standards for treatment facilities, new landfills, and a national manifest system; and the construction of several new incinerators, including three large ones for PCBs and other organics, and the construction of a mobile incinerator (EPA) for site clean-up.

Things not done well include: the development of a national program whereby all waste is regulated; the full regulation of problem technologies; the development of strong enforcement/compliance programs at the state level, to include adequate personnel and training and consistency in standards and policy priorities; the development of permitting capability that would have allowed approval of more than the 500 of the 5000 facility applications received; development of programs to deal effectively with groundwater pollution at existing facilities and abandoned sites; and the development of a national system whereby landfilling, injection well disposal, and lagoons did not predominate, with less than 1% of hazardous waste being treated.

The following areas are in need of attention: the gaps in legislation pertaining to waste oils, small generators, organics from chemical and petroleum industries and the mining industry; programs to deal with potential problem areas such as municipal and non-hazardous industrial landfills and waste water lagoons; additional control of underground tanks; and upgrading of enforcement to expedite permitting and the construction of new treatment facilities.